THE PHILOSOPHY OF MEDICINE

Framing the Field

Philosophy and Medicine

VOLUME 64

The titles published in this series are listed at the end of this volume.

THE PHILOSOPHY OF MEDICINE

Framing the Field

Edited by

H. TRISTRAM ENGELHARDT, JR.

Philosophy Department, Rice University,
Baylor College of Medicine

KLUWER ACADEMIC PUBLISHERS
DORDRECHT / BOSTON / LONDON

A C.I.P. Catalogue record for this book is available from the Library of Congress.

ISBN 978-90-481-5419-7 e-ISBN 978-0-306-47475-0

Published by Kluwer Academic Publishers,
P.O. Box 17, 3300 AA Dordrecht, The Netherlands.

Sold and distributed in North, Central and South America
by Kluwer Academic Publishers,
P.O. Box 358, Accord Station, Hingham, MA 02018-0358, U.S.A.

In all other countries, sold and distributed
by Kluwer Academic Publishers, Distribution Center,
P.O. Box 322, 3300 AH Dordrecht, The Netherlands.

Printed on acid-free paper

Printed in the Netherlands.

STUART FRANCIS SPICKER, Ph.D.
B.A., City University of New York, Queens College, 1959
M.A., New School for Social Research, New York City, 1962
Ph.D., University of Colorado, Boulder, 1968

Professor
Massachusetts College of Pharmacy and Health Sciences
Boston, Massachusetts

Professor Emeritus
University of Connecticut Health Center
Farmington, Connecticut

TABLE OF CONTENTS

H. TRISTRAM ENGELHARDT, JR.

THE PHILOSOPHY OF MEDICINE AND BIOETHICS: AN INTRODUCTION TO THE FRAMING OF A FIELD

I. AT THE ORIGINS OF A FIELD

From where do new fields of scholarship arise? How do they take shape? Why do old ones reappear and gain new life? In particular, how and why in the 1970's did the philosophy of medicine again become an important field of study? Why did philosophy become so central to the new and emerging field of bioethics? There are no simple answers to these questions. Cultural developments represent an interplay of values, institutions, science, technology, economic forces, and, most importantly, persons. In the 1960's and early 1970's, the number of persons who credited the importance of the philosophy of medicine and bioethics could have easily assembled in a small room. Of these, few had an appreciation of the breadth of philosophical issues at stake in medicine and the biomedical sciences. Stuart Spicker did. Fields take shape in great measure because persons see their possibility and have the insight to realize them in actuality. Stuart Spicker's vision of bioethics and the philosophy of medicine has had an enduring impact on this field.

This is not to deny the role of ideas. Ideas act as final causes for fields of scholarship. They inspire scholars who call into existence the social structures within which actual scholarly dialogue can occur. There are no scholarly fields without discussions, conferences, journals, and books. There are no conferences, journals, or books without scholars to bring them into being. Organizing conferences, establishing journals, and developing book series are labors that require energy, insight, and the ability to stick with tasks over time. This has proved no less true with bioethics and the philosophy of medicine. These areas of interest formed a scholarly field united by the interest of philosophers in medicine and the biomedical sciences. It came to be appreciated that, with respect to medicine and the biomedical sciences, issues of evaluation and explanation were intertwined: a distinction between the philosophy of medicine and bioethics was not only artificial but misleading. The two areas of concern came to function as one general field of investigation with areas of special focus. This field came into existence because certain

H. T. Engelhardt, Jr. (ed.), The Philosophy of Medicine, 1–15.

people saw its importance and were willing to labor patiently to establish a full-fledged field of academic pursuit.

At the end of the 1960's and the beginning of the 1970's, conferences had begun to recognize puzzles of a philosophical sort about medicine. First, there was a cultural hunger to locate medicine within larger cultural concerns. This was driven by the increasing salience of medicine itself. In every developed country, health care was claiming a larger portion of the gross domestic product. The promise of new technologies also threatened seemingly novel moral and public policy puzzles. If the law deserved a philosophy of law, medicine seemed to deserve something like a philosophy of medicine. Although that term was hardly used in the 1960's (Szumowski, 1949), philosophy of medicine had a history that reached to the beginning of the 19th century (Berlinghieri, 1801). Second, new technologies such as critical care units and organ transplantation pressed for clarity about issues such as determining when death occurs. Third, old moral problems became more acute because the technologies that occasioned them had become safer and more tempting: for example, abortion became associated with fewer risks of morbidity and mortality, and there was the promise of ever more effective prenatal diagnosis and selective destruction of fetuses. Fourth, there appeared to be purely philosophical issues such as the nature of disease and illness that were addressed by neither the philosophy of medicine nor even the philosophy of biology. There were these and many other issues of evaluation and explanation associated with medicine that have since given birth to sustained philosophical reflection and analysis. Few philosophers seemed to notice them. Or if they noticed them, they were not yet sure how to name them or whether to place them within an independent field.

There was even a resistance to taking the philosophy of medicine and bioethics seriously as fields for philosophical exploration (Engelhardt and Spicker, 1975, pp. 211–234). Philosophers were disinclined to entertain this area of scholarship. It appeared to them to involve "only" issues of application. Physicians were resistant to the view that philosophers could contribute to the understanding of their profession. Actual clinical experience alone was taken to provide the appropriate insight into medical moral matters. Then things began to change, in great measure because of the insight, intelligence, and energies of a small number of individuals such as Andre Hellegers, Director of the Kennedy Center, Willard Gaylin and Daniel Callahan, President and Director of the Hastings Center, Chester Burns, Director of the Institute for Medical Humanities, and Edmund D. Pellegrino, Chairman of the Institute of Human Values in Medicine. In 1971, the Hastings Center Report came into print. Then in 1976 the *Journal of Medicine and Philosophy* published its first issue under the editorship of Edmund Pellegrino and with the support of the Society for Health and Human Values. From the Kennedy Institute through Warren Reich, the *Encyclopedia of Bioethics* appeared in 1978 (Reich, 1978). There were other institutions and publications that framed the field. Among them must surely be counted the book series within which this volume appears: *Philosophy and Medicine.* For a quarter of a century and for over sixty volumes, the series had Stuart F. Spicker (1937–) as its co-editor. Though mentioned only in passing in Jonsen's excellent study of the birth of bioethics (Jonsen, 1998), the re-emergence of interest in the philosophy of medicine had a significant impact on the philosophical character of the field of bioethics. In the above-mentioned list of founding fathers of the new field, only Callahan and Spicker are philosophers *sensu stricto.*

Spicker is one of the few who at the beginning of the 1970's recognized the importance of the philosophy of medicine in its own right and its significance for bioethics in particular. When he first turned to bringing the energies of philosophy to medicine and the biomedical sciences, one could find few individuals at professional philosophical meetings to take this project seriously. Yet, Stuart Spicker recognized that philosophers had to be brought to engage the field of medicine. In the absence of philosophy's discipline, the bioethics that was emerging would be conceptually shallow and important issues would be inadequately addressed. Spicker recognized that medicine, the biomedical sciences, and bioethics would not be appropriately treated unless serious philosophical scholarship addressed the foundations of bioethics and the philosophy of medicine. In addition, he correctly saw that such examinations needed a substantial vehicle for their expression. Stuart Spicker was convinced that there must be scholarly conferences to produce a literature and then there must be vehicles for its publication. This insight led to the establishment of the *Philosophy and Medicine* series.

Stuart F. Spicker joined in launching the *Philosophy and Medicine* series in a period when the term "bioethics" had just been coined.[1] There was little anticipation of the breadth and depth of the intellectual interest that would come to support this area of scholarship. There was much foment about the dehumanizing character of technology and the new medicine. Many were reflecting on the values at stake in new scientific and technological developments. Few, however, appreciated the ways in which this period was both novel and had a continuity with past scholarly undertakings. Moral reflections concerning medicine had a substantial theological tradition of substantive literature extending to the beginning of the 16th century and the prominence of the new sciences (Cronin, 1958). This scholarly tradition, though it had a substantive philosophical component, had its roots severed in the spiritual chaos that followed the Second Vatican Council. Roman Catholics were no longer interested in carrying it further (Kelly, 1979). The emerging secular culture was losing its interest in religious medical-moral reflections, especially those with a sectarian character. Medical ethics, which had gained a prominence at the end of the 18th and the beginning of the 19th century (McCullough, 1998a, 1998b), was about to be transformed and relocated by a new field: bioethics.

This relocation and the rebaptism of medical ethics under the term bioethics was of broad as well as profound significance. It was broad in bringing the internal values of medicine and the health professions within general cultural concerns, a point made by Henk ten Have in his essay in this volume. This relocation was of profound significance because it was integral to the emergence of a post-traditional society, which was deprofessionalizing medicine. Medicine was no longer to be an independent guild. It was no longer a self-regulating profession. Medicine was being recast both as a trade and as a governmentally controlled societal undertaking. As a consequence, medicine lost much of its professional independence. Very quickly, only a few physicians remained independent professionals. All would find themselves in a less independent profession. This radical transformation of medicine was integral to the emergence of a post-traditional understanding of social structures and authority, requiring medical ethics to be subordinate to a field of moral vision independent of the health professions and open to public scrutiny.

This transition was disorienting to physicians. In great measure, they never fully appreciated what was taking place. Patients for their part had changed their roles: they had begun to demand more rights over against both physicians and their own families. Many patients wanted to be treated as independent, isolated decision-makers. Patients, too, only imperfectly recognized that they were caught up in a major social change in Western culture's self-understanding. The Christian assumptions that had framed Western Europe and North America were in decay, and a new secular culture was emerging with its new experts and its new, albeit secular, clergy (Engelhardt, 2000). One may think here of the entrance of ethicists as special secular chaplains. George Agich in his essay in this volume reflects this transformation as he explores the roles played by ethicists, roles that had once self-evidently been played by rabbis, priests, and ministers. In this period of change and disorientation, there was a hunger for value and moral substance. The result was a striking resurgence of interest in the humanities and their connection with health care. In the absence of religious norms, many turned to the humanities for guidance.

In many respects, this resurgence of the humanities in medicine had roots in the New Humanism and the Third Humanism, which had developed at the beginning of the 20th century with such figures as Abraham Flexner (1866–1959) and Werner Jaeger (1888–1961). Abraham Flexner, who had played such an important role in the reform of medical education at the beginning of the 20th century, contributed as well to a recognition of the role of the humanities in society (Flexner, 1910, 1928). Werner Jaeger through his work (Jaeger, 1943–45) became an author esteemed by the most important figure in the 1970's concerned with the medical humanities: Edmund Pellegrino. Pellegrino shaped an understanding of the intimate connection of the humanities in medicine through his famous and insightful observation, "Medicine is the most humane of sciences, the most empiric of arts, and the most scientific of humanities" (Pellegrino, 1979, p. 17). Stuart Spicker recognized the richness of this observation before it had been made. Spicker also saw something that many in the medical humanities had never appreciated. He understood the need to narrow the focus from the medical humanities in general to that area of the humanities that could analytically assess the expansive claims being raised and the vague questions being forwarded: philosophy. As he stressed, the medical humanities were not one field, but a collage of heterogeneous fields: philosophy of medicine, bioethics, the history of medicine, literature in medicine, religion in medicine, law in medicine, etc. What was needed was a systematic, critical, indeed philosophical investigation of what was at stake. Spicker put emphasis on the philosophy of medicine, including a philosophical bioethics, as the necessary condition for the possibility of giving coherence to the medical humanities. Spicker defended the importance of philosophy and the importance of involving philosophers in the emerging field.

This volume is a Festschrift celebrating this insight. It marks twenty-five years of Stuart Spicker's creative labors with the *Philosophy and Medicine* series. It is also a work directed to a field that took shape through the energies and insights of the honoree. The *Philosophy and Medicine* series is his greatest work. Though the particular essays often mention particular contributions by Spicker to the literature, they are all involved with Stuart Spicker's scholarship through the *Philosophy and Medicine* series, which he brought into existence. They have all been influenced by him. Stuart Spicker in great measure has done for the field what a good dean does for a university. He envisaged how creative scholarship should be undertaken. He raised money and invested his energies so that opportunities could be transformed into resources nurturing a growing field. Having a keen sense of people and an ability to engage scholars of wit and inventiveness, he brought together men and women with interest and ability to explore the foundations of a developing area of scholarship. Very importantly, Stuart Spicker began to look not primarily for those who were already interested in bioethics and the philosophy of medicine, but for those who he thought should become interested and whose reflection could enrich the field. He looked for people who could see things anew. It was clear to Spicker that the worst way to nurture a nascent field of scholarship would be to envisage it as a social crusade or in terms of well-worn expectations from other fields. Spicker knew that it was best to find people who were able creatively to go against established expectations, framing issues from a new perspective, while at the same time placing the novel within the history of ideas. He brought to the symposium series that initially supported this series of publications philosophers who could be engaged to look at bioethics and the philosophy of medicine creatively and with a critical eye. These included persons such as Marjorie Grene, R. M. Hare, Patrick Heelan, Alasdair MacIntyre, Stephen Toulmin, Marx Wartofsky, and Richard M. Zaner.

From the outset, Spicker also understood that the field would be enriched only if it incorporated scholarship and reflection from Continental as well as analytic philosophy. His vision has remained broad. Through his appreciation for Continental philosophy, Spicker nurtured a bridge between American and European philosophical approaches to the biomedical sciences and health care. This was important in that European philosophical concerns with medicine have maintained an interest in models of explanation and philosophical anthropology that have located bioethics within broader philosophical projects. Spicker helped take the first important steps to make bioethics and the philosophy of medicine more than an American phenomenon and to draw upon the substantial concerns and literature within Continental philosophy. It was thus quite natural that the *Philosophy and Medicine* series brought to publication volumes that affirmed continuity with the scholarship of the 18th and early 19th centuries (Löwy, 1990). Although many overlooked the circumstance that Thomas Kuhn's widely influential *The Structure of Scientific Revolutions* (Kuhn, 1962) had deep resonances with a brilliant earlier book in the philosophy of medicine (Fleck, 1935), the *Philosophy and Medicine* series took these roots quite seriously (Delkeskamp-Hayes and Cutter, 1993).

Because of his broad appreciation of issues in philosophy and his ability to work across philosophical boundaries, Spicker was able to bring together scholars from Europe, China, Israel, Japan, and North America to work in the philosophy of medicine and in philosophical approaches to bioethics. Out of this rich appreciation of what was necessary to give a field shape, Stuart Spicker created scholarly frameworks for publication in which an international area of reflection and analysis could mature. The field of the philosophy of medicine and bioethics would literally not be such as it is, had not Stuart Spicker given to this field as he has.

II. GENEROSITY, COLLEGIALITY, AND FRIENDSHIP

It was at a meeting of the American Philosophical Association, Eastern Division, in Boston in December, 1972, that I first met Stuart Spicker. Just that year I had graduated from medical school and had in September joined what was to become the Institute for the Medical Humanities at the University of Texas Medical Branch in Galveston. I was publishing as much on the history of medicine as on the philosophy of medicine. Sometime during my last year in medical school, I had sent a letter to André Hellegers after reading about the founding of the Kennedy Institute in 1971. I had received no reply. I was looking for colleagues, for those who saw the promise of taking seriously as a philosopher the issues raised by medicine. My work to that date had been primarily focused on Kant, Hegel, and Husserl, as well as on certain elements of the history of medicine in the 18th and 19th centuries. Because of all of this, and because of the upcoming publication of my first book, *Mind-Body: A*

Categorial Relation (Engelhardt, 1973) and my translation with Richard Zaner of *The Structures of the Life-World* (Schutz and Luckmann, 1973), I gravitated to those in the so-called Continental tradition. Thus, I came to be introduced to Stuart Spicker, Marx Wartofsky, and Robert Cohen. Stuart Spicker was a joy, a kindred soul, a philosopher who understood that any attempt to approach issues in the philosophy of medicine and bioethics should be tied to the history of thought and the broad concerns of philosophers. Out of these discussions, it became only too natural that we considered establishing a series on the model of *Boston Studies in the Philosophy of Science*. Spicker promised to do the legwork. Over the next few months, Stuart Spicker remarkably brought together the commitment of Dr. Reidel and his company to the publication of the new series.

Spicker also had the wit, wisdom, and understanding to recognize not just that this field could take form and focus only if scholars could be brought into substantial dialogue, and if there were vehicles available to sustain the emerging scholarship. He also had the talent to create the actual structures needed. It was Spicker who went to Dr. D. Reidel to secure the series contract for *Philosophy and Medicine*. With the help of Robert Cohen and the late Marx Wartofsky, they together made it happen. With Spicker's reflections and labor, the First Trans-Disciplinary Symposium on Philosophy and Medicine was held in Galveston, Texas, May 9–11, 1974. From this symposium, a new book series achieved reality. Reidel, and its successor publisher Kluwer Academic Publishers, never lost sight of the importance of this undertaking. Because of the presence of Stuart Spicker and his vision of scholarship, Jan Hattink, Alexander Schimmelpenninck, the late Mrs. Nel Jones (Acquisitions Editor), Joop Wibier (Desk Editor), Patrick Wharton, Gerald Curtis, Maja de Keijzer, H. E. Zoeteman, Anny Burer, Jan-Hendrik van Leusen, Stephanie Harmon, and many others have collaborated in a book series now spanning over 60 volumes. Stuart also helped in the development of other publications. He joined the Editorial Board of the *Journal of Medicine and Philosophy*, later to become its Associate Editor. Subsequently, he became co-editor of the book series *Clinical Medical Ethics* and editor of the journal *Health Care Ethics Committee Forum*. Many of the major scholarly publications in this area of scholarship would not have their character without Spicker's contributions.

Spicker began to plan a series of symposia that would critically assess the possibilities of the new field. Putting together conferences requires collaboration. Bringing people together at a conference that will break new ground requires engaging the participants one with another around a common task. This is a work of art. In the process, one comes to see the character of all involved, their strengths and their weaknesses. Over the years, as more symposia were held and more volumes came to print, I discovered again and again the remarkable generosity of Stuart Spicker. He showed himself to be generous twice over. First, he was generous in reaching out in kindness to others, considering what would please others, not himself. He gives from the heart. But second and more importantly, in a world often marked by small-mindedness and avid self-seeking, Stuart showed himself able to overlook slights, to give a beneficent interpretation to what may have been malevolent, and with kindness to work as far as possible with individuals who at times were disinclined to collaborate. In a world where the motto is often "don't get mad, get even," Stuart has striven to forget vexations in order to pursue larger goals of scholarship and the very humane goal of collegial conversation. From his blend of generosity and understanding of people, Spicker has again and again brought peace when people were splenetic about some element of a symposium or a detail of publication. In ways that defied the expectations of the circumstance, Spicker assuaged bruised egos and enabled things to go forward.

Many of the volumes of the *Philosophy and Medicine* series, and nearly all of the first ones, took shape through a dialogue in which Stuart Spicker would match persons with ideas. The solution to these most difficult of puzzles was often found around very good food and distinguished but reasonably priced wine in obscure restaurants across the globe. Here, Stuart has an unfailing talent. Whether in New York or Strasbourg, Madrid or Tel Aviv, he has a nose for which restaurant will marry a good chef with a good wine cellar. Though in other regards Stuart Spicker gives heavy emphasis to philosophical method over the broad and general concerns of the humanist, here Stuart Spicker has revealed himself to be the humanist par excellence. He has the insight of the cultivated Roman, the *humanissimus vir*: intellectual creativity is nurtured through the joys of our culture. Though the phrase *in vino veritas* is usually engaged with quite a different meaning, Spicker knew that a good wine can be a substantive support of creative dialogue in the pursuit of philosophical truth. It is a truth of humanities that the celebration of a good meal with fine wine among friends can be the basis of a fruitful symposium. The celebration of food, wine, and ideas can with friendship open a dialogue that journeys without prejudice into unanticipated areas of taken-for-granted assumptions and the critical assessment of standing convictions. A meal with Stuart over good spirits would give birth to the vision for symposia breaking new ground, and thus to volumes that would change the character of scholarly debates. The success of the *Philosophy and Medicine* series is indebted to Spicker's humane generosity and his appreciation of a life lived *humaniter*.

It is because of Spicker's attention not just to ideas but to persons in the grace of good conversation that I believe his work has been so well received in Japan, a culture that has traditionally esteemed harmony in collaboration with colleagues. Japanese culture acknowledges that important scholarly collaboration does not occur merely in lecture halls or in seminar rooms, but also around good wine, drink, and conviviality. Independently of a knowledge of Plato's *Symposium*, there is an appreciation that ideas are nurtured by persons who have a broad and felt comprehension of the interplay of intellectual concerns and the commitments of culture. This is also an insight at the core of Spicker's being. At every juncture, Stuart has been able to speak across cultural barriers and to collaborate with scholars of different philosophical commitments. This book series would never have had its character without this strength of Spicker's character.

III. A DIVERSITY OF VOICES

Stuart's broad appreciation of the philosophy of medicine and bioethics is reflected in this volume's essays. Their methodology and content testify to the scope of approaches that have shaped the field over the quarter-century since the first Trans-Disciplinary Symposium on Philosophy and Medicine in 1974. The diversity of their methods speaks to the breadth of Spicker's influence.

The first cluster of essays draws on a major area of Stuart Spicker's work: the philosophy of the body and its importance for an adequate anthropology. The first essay by Henk ten Have employs Stuart Spicker's insights regarding the crucial role of philosophical anthropological phenomena in understanding the body as it appears for medicine. Medicine is unique, as ten Have emphasizes, because it treats subjects who *have* a physical body and *are* a lived body. Spicker and ten Have recognize that a new and richer conception of bioethics is emerging, which in part is tied to the changed social status of medicine, and in part to a more philosophical, indeed anthropological understanding of the key elements of bioethics. This conception is integral to the realization that moral problems require the context of a philosophical anthropology. An adequate bioethics demands a philosophy of the body and a philosophical anthropology (Leder, 1992).

The next essay is by a brilliant and faithful friend, now deceased, who again and again contributed a fresh vision to the philosophy of medicine: Marx Wartofsky. At the symposium producing the first volume in this series, it seemed as if Wartofsky would not make it to the podium in time for his presentation. The audience was waiting, the speaker was not present, he seemed to have disappeared. At the very last minute Wartofsky strode in, relaxed, self-possessed, his hair still wet from the water of the Galveston littoral. His presentation was superb. With that paper, "Organs, Organisms and Disease: Human Ontology and Medical Practice" (Wartofsky, 1975), he began the first of a series of contributions to the volumes of this series, the last of which is published here. In this essay he investigates human identity and the interplay between wholes and parts, issues that resonate with the scholarship of both Spicker and ten Have. Wartofsky redirects us to a theme important in Spicker's work, namely, that one must appreciate the different levels of discourse involved in medical explanation. Finally, the essay by Mark Cherry addresses the body as a source of organs. Here, crucial issues surface about the objectification and commodification of bodies and their parts. Once again a theme from Spicker's work becomes central: the human body is also a lived body. How should this bear on the proper treatment of organs and their exchange in markets? Each of these essays vindicates Stuart Spicker's concern that a phenomenology of the body and the philosophical exploration of its significance in medicine is core to any successful account of bioethics.

The next cluster of essays reflects the radical transformation of moral and social expectations at the root of the emergence of secular bioethics and the rebirth of the philosophy of medicine in the 1960's and 1970's. Thomas Halper offers a careful exploration of Holland's development of a permissive health care policy with regard to physician-assisted suicide and euthanasia, while conservatively attempting to avoid definitive legislation or judicial decisions. In providing an account of the Dutch practice of euthanasia, Halper offers a picture of the transformation of

The Netherlands into a post-Christian society in which the new secular bioethics can find ever less place for traditional insights into the evil of physician-assisted suicide or euthanasia. The result is a policy in many ways paternalistic, as Halper observes, yet in truth, radically new. Set against legislation elsewhere in the world, it is a significant exemplar of a sea change in Western cultural values. Many of the volumes in the *Philosophy and Medicine* series have also documented this collapse of traditional moral assumptions and the movement of secular bioethics into the resulting cultural and moral vacuum (e.g., Spicker *et al.*, 1987).

A similar report on a radical change disguised as a mere contribution to better health care decision-making is provided by George Agich's essay on the role of ethics consultants. In comparing the role of ethics consultants to the roles of policemen and parents, Agich offers *en passant* a further exploration of the post-traditional character of society in which parents have authority no longer on the basis of status,[2] but from the performance of an important task (i.e., raising a child). In this model, ethicists emerge as the new secular chaplains unordained and therefore not in authority. They gain authority as they aid puzzled patients, physicians, and other health care professionals to make decisions in a post-traditional, post-Christian, post-professional world. It is after all this new social world that engendered the hunger for a new discipline and the ethicists it would beget. In a similar vein, Loretta Kopelman provides a critical exploration of the paternalistic role of drug licensing authorities and the demand of patients to be able to make use of treatments before their safety and efficacy has been fully assessed. In a social world where authority is most plausibly drawn from the consent of the governed, a government's paternalism in protecting persons individually and as a society from possibly inefficacious and harmful drugs for a disease that is terminal becomes ever less plausible. The pursuit of social utility through well-designed research protocols may obscure the circumstance that those enrolled in such therapeutic research trials may receive optimal treatment. Individuals may also be unwilling to support the advancement of knowledge about treatment, if this sets their individual welfare at risk. Patients may not be willing to defer treatment in the service of well-designed comparisons of possible therapeutic approaches. A liberal, if not libertarian, ethic by default becomes the framing morality of secular bioethics.

The final cluster of papers examines key issues and categories in the philosophy of medicine. William Fulford devotes his energies to the philosophy of psychiatry, a point of considerable interest to Stuart Spicker and an area in which he has published importantly. The essay bears out the insights of Spicker that psychiatry must be understood broadly and that a narrow appreciation of psychiatry would misguide the clinician and harm the patient. The reorientation of concerns with equity and equality in the face of urbanization and the technological transformation of society is addressed by Bernard Dickens. This essay shows the dynamic interplay of ideas and technologies, a basic realization that Stuart Spicker brought to the framing of the *Philosophy and Medicine* series. The next essay locates bioethics within concerns regarding the state and social structure. In his essay, Thomas J. Bole, III, through his exploration of Immanuel Kant and G. W. F. Hegel, demonstrates how tightly the philosophy of medicine and bioethics are bound to traditional philosophical issues that set the framework for our contemporary understanding of bioethics. He shows how the work of these figures is integrated with our current concerns. Though I may disagree significantly with Bole about how to read these figures (see my 'Bole on Kant, Hegel and Engelhardt: A brief reply,' following his essay), it is clear that attending to their work will bring us closer to understanding the possibilities for a secular bioethics. Finally, Akio Sakai addresses a theme to which Stuart Spicker has given much reflection: the significance of history and of time. This short essay reminds us that the philosophy of medicine relocates bioethics in the broad understandings that frame our cultures.

The philosophy of medicine is more than the bioethics of hospital ethicists, though Spicker understands well both this bioethics and the contemporary role of ethicists in hospitals. The philosophy of medicine is more than an important analytic study of issues such as abortion, paternalism, and death, though such studies have given precision to bioethical debates. The philosophy of medicine is even more than a marriage of Continental and analytic philosophical projects for the exploration of medicine and the biomedical sciences. The philosophy of medicine is also a systematic and critical dialogue born of wonder and agony concerning human disease, illness, defect, and disorder, as well as the ways in which they can be treated and understood. As such, the history of the philosophy of medicine is not just a history of philosophies, but of philosophers who in particular times and at certain places give philosophy its incarnation. Stuart Spicker is one of those people in whose absence the philosophy of medicine and bioethics would have been otherwise, and in that circumstance impoverished.

Center for Medical Ethics and Health Policy
Baylor College of Medicine/Rice University
Houston, Texas

NOTES

[1] Though it is most likely that the term bioethics was first coined by Van Rensselaer Potter in 1970 (Potter, 1970, 1971), he intended the term to identify a new global ethos built around a set of environmental concerns. His meaning of bioethics is not the one that defined the field. That term was given its contemporary meaning by André Hellegers, even if not independently created by him (Reich, 1994).

[2] Parents being in authority over their children as a matter of status would include their having authority because they produced their children. Such parental authority would govern children, unless defeated by particular circumstances such as abandonment of the child, placing the child for adoption, child abuse, failure to support the child, etc.

BIBLIOGRAPHY

Berlinghieri, F. V.: 1801, *La Filosofia della Medicina*, aux frais de l'auteur, Lucca.

Cronin, D. A.: 1958, *The Moral Law in Regard to the Ordinary and Extraordinary Means of Conserving Life*, Ph.D. dissertation, Gregorian University, Rome.

Delkeskamp-Hayes, C. and Gardell Cutter, M. A. (eds.): 1993, *Science, Technology, and the Art of Medicine*, Kluwer, Dordrecht.

Engelhardt, H. T., Jr.: 1973, *Mind-Body: A Categorial Relation*, Martinus Nijhoff, The Hague.

Engelhardt, H. T., Jr. and Spicker, S. F. (eds.): 1975, *Evaluation and Explanation in the Biomedical Sciences*, D. Reidel, Dordrecht.

Engelhardt, H. T., Jr.: 1996, *The Foundations of Bioethics*, 2nd ed., Oxford University Press, New York.

Engelhardt, H. T., Jr.: 2000, *The Foundations of Christian Bioethics*, Swets and Zeitlinger Publishers, Lisse.

Fleck, L.: 1935, *Entstehung und Entwicklung einer wissenschaftlichen Tatsache*, Benno Schwabe, Basel.

Flexner, A.: 1910, *Medical Education in the United States and Canada, A Report to the Carnegie Foundation for the Advancement of Teaching*, Bulletin No. 4, Carnegie Foundation, New York.

Flexner, A.: 1928, *The Burden of Humanism*, Clarendon Press, Oxford.

Jaeger, W.: 1943–45, *Paideia: The Ideals of Greek Culture*, Oxford University Press, Oxford.

Jonsen, A. R.: 1998, *The Birth of Bioethics*, Oxford University Press, New York.

Kelly, D. F.: 1979, *The Emergence of Roman Catholic Medical Ethics in North America*, Edwin Mellen, New York.

Kuhn, T. S.: 1962, *The Structure of Scientific Revolutions*, University of Chicago Press, Chicago.

Leder, D. (ed.): 1992, *The Body in Medical Thought and Practice*, Kluwer, Dordrecht.

Leoni, B.: 1991, *Freedom and the Law*, 3rd ed., Liberty Press, Indianapolis, Indiana.

Löwy, I.: 1990, *The Polish School of Philosophy of Medicine*, Kluwer, Dordrecht.

McCullough, L.: 1998a, *John Gregory and the Invention of Professional Medical Ethics and the Profession of Medicine*, Kluwer, Dordrecht.

McCullough, L. (ed.): 1998b, *John Gregory's Writings on Medical Ethics and Philosophy of Medicine*, Kluwer, Dordrecht.

Nozick, R.: 1974, *Anarchy, State, and Utopia*, Basic Books, New York.

Pellegrino, E.: 1979, *Humanism and the Physician*, University of Tennessee Press, Knoxville.

Potter, V. R.: 1970, 'Bioethics, the Science of Survival,' *Perspectives in Biology and Medicine* 14, 127–153.

Potter, V. R.: 1971, *Bioethics, Bridge to the Future*, Prentice-Hall, Englewood Cliffs, NJ.

Reich, W. (ed.): 1978, *Encyclopedia of Bioethics*, Macmillan Free Press, New York.

Reich, W.: 1994, 'The Word "Bioethics:" Its Birth and the Legacies of Those Who Shaped Its Meaning', *Kennedy Institute of Ethics Journal* 4, 319–336.

Schutz, A. and Luckmann, T.: 1973, *The Structures of the Life-World*, R. Zaner and H. T. Engelhardt Jr. (trans.), Northwestern University Press, Evanston, Illinois.

Spicker, S. F., Bondeson, W. B., and Engelhardt, H. T., Jr. (eds.): 1987, *The Contraceptive Ethos*, D. Reidel Publishers, Dordrecht.

Szumowski, W.: 1949, 'La Philosophie de la Médecine, son histoire, son essence, sa dénomination et sa définition,' *Archives Internationales d'Histoire des Sciences* 9, 1097–1141.

Wartofsky, M.: 1975, 'Organs, Organisms and Disease: Human Ontology and Medical Practice', in H. T. Engelhardt Jr. and S. F. Spicker (eds.), *Evaluation and Explanation in the Biomedical Sciences*, D. Reidel Publishers, Dordrecht, pp. 67–83.

SECTION ONE

THE PHILOSOPHY OF THE BODY AND BIOETHICS

HENK A. M. J. TEN HAVE

BODIES OF KNOWLEDGE, PHILOSOPHICAL ANTHROPOLOGY, AND PHILOSOPHY OF MEDICINE

I. INTRODUCTION

In the second issue of the *Journal of Medicine and Philosophy* Stuart Spicker analyzes the possibility of a philosophy of medicine (Spicker, 1976). He argues that philosophy of medicine as a new subdiscipline exists if and only if the Cartesian notion of "body" is abandoned, the physical body which is described in the language of things and explained as a physiological mechanism. As long as we focus on the complexity of the physical, anatomical and physiological body, we already have a philosophy of biology to attend to the problems germane to living bodies. Philosophy of medicine only comes into existence, in the view of Spicker, once we "acknowledge . . . the lived body" (Spicker, 1976, p. 128). Medicine is unique in that it deals with subjects who *have* at the same time a single, physical body and *are* a lived body. Problems arise because of medicine's tendency to construe the patient's lived body as a physical body.

Re-emphasizing the significance of the lived body as the unique and proper domain of medicine, Spicker during the modern revival of philosophy of medicine consistently draws attention to the importance of philosophical anthropology (Spicker, 1975; 1986; 1990). However, the relationship between philosophy of medicine and philosophical anthropology is not clear. Spicker develops a specific anthropological view: the human condition has at least one fundamental structure, *viz.*, that it is open to infirmity and instability. Crucial is the image of man as infirma species. Exploring the notion of infirmity and therefore developing a philosophical anthropology is then a precondition for the emergence of philosophy of medicine as a new subdiscipline. Medical philosophical anthropology is an a priori discipline. Spicker argues that it should be distinguished from philosophy of medicine, since the latter analyzes the presuppositions of medicine whereas anthropology focuses on the explication of basic human structures. In the same publication, he states that the notion of "infirmity" is one concept within the philosophy of medicine (Spicker, 1976, p. 129); when exploring such notion is the

H. T. Engelhardt, Jr. (ed.), The Philosophy of Medicine, 19–36.

objective of philosophical anthropology, it is already included in philosophy of medicine.

This last view is consonant with the development of medical philosophy over the last century. In the recent history of philosophy of medicine, examination of the foundations of medicine has led to an anthropological tradition, particularly in Germany and the Netherlands. This tradition has declined with the general and rapid growth of interest in moral issues over the last three decades. Recently, patients, philosophers and physicians are becoming more aware that many moral problems cannot be resolved without clarification and interpretation of the images of man that underlie ethical perplexities. Such awareness has led to a renewed interest in the writings and ideas of anthropologically oriented scholars of the first half of this century. Since they try to change medical practice and theory 'from within', connecting daily experiences with philosophical insights, and delineating a renovated science of the human person, they have in fact the same motivations and values as many scholars interested in medical ethics and medical philosophy today. Stuart Spicker's reminder of the significance of philosophical anthropology twenty years ago is therefore valid even today, when philosophy of medicine is so much dominated by bioethical activities.

II. MEDICINE'S RELATION TO PHILOSOPHY

In its history, medicine has maintained a long-standing dialectical relation with philosophy (ten Have, 1980; 1983). Medicine apparently has evolved concomitantly with philosophy of medicine. To define the identity of medicine and to demarcate medicine as an autonomous discipline, authors of medical treatises used to dispute the contemporary speculations about the nature of the world.

The philosophical conception of order in nature, for example, was useful to demarcate medicine from mythology and religion, postulating natural disease causation and regularity in the natural history of diseases. The confrontation with philosophy led these physicians to discover the relevance of experience and precise observation for the practice of medicine. The emancipation of medicine as a discipline separated from philosophy arose therefore not from a rupture with philosophy but from a philosophical re-interpretation of what is characteristic of medicine itself.

However, when medicine had become firmly established as a unified science of disease, using the methodlogy of the natural sciences as the basis for obtaining and applying knowledge, a fundamental change of the relationship with philosophy occurred. Philosophy is no longer in competition with medicine, trying itself to understand and explain the phenomena of illness, suffering, incapacity, and death, but philosophical reflection is taking the medical approach of these phenomena as its object of analysis. This is an important shift in level of reflection and study. Instead of operating at the same conceptual level as medicine and examining the same objects, philosophy has moved to a meta-level, analyzing medicine as its proper object. In the emancipatory process of individual sciences for some time a speculative, philosophical approach has existed alongside of empirico-experimental approaches. Well-known examples are the medical systems propounded by idealist philosophers in nineteenth century Germany (Risse, 1976). These systems of philosophical medicine were felt to be competitive with the new ethos of medicine as a natural science. The long-standing and intimate bond between medicine and philosophy was therefore re-interpreted in the last century as an antagonistic relationship. Many clashes between the two disciplines arose. The conflict was solved by re-defining the role and object of philosophy *vis-à-vis* medicine. Philosophy was re-located as a science of science at another level of abstraction, not concerned any longer with medicine's object of study and intervention but with the concepts and methods employed by medicine in studying and manipulating its objects. The outcome of this re-location was the emergence of philosophy of medicine as a separate intellectual discipline. Conceiving philosophy as a meta-activity, a competitor was turned into a critical ally.

The rise of philosophy of medicine in the nineteenth century was connected with transformations in the social status and the self-image of medicine which were settled in most countries in the second part of the century. Until that time medicine was practiced by many kinds of healers. University-trained physicians were a minority. The heterogeneity of medical practitioners and their constant struggle over competencies and privileges gave cause for many satire, ridiculing the doctors' ignorance and false pretensions. Medicine was also associated with poor status and low income. The nineteenth century process of professionalization reversed this situation. One of the claims promoting the creation of an unified and organized medical profession was that medicine could be developed and practiced upon a scientific foundation superior to that of other healers. This claim generated a renewed interest in philosophical reflection. Professional cohesion and uniformity of medical practice, brought about by external forces, are in need of internal justification; they motivate a search for the identity of the now unified discipline. The central question guiding philosophy of medicine at that time was: What are the specific characteristics of medicine?

Elsewhere I have argued that in the thematic development of philosophy of medicine over the last hundred years, particularly in Germany, France and Poland, three traditions can be distinguished: an epistemological, anthropological and ethical tradition (ten Have, 1996).

Flourishing around the beginning of the last century was an epistemological tradition. Medicine is here characterized as primarily a natural science. Essential are the methods of acquiring and applying knowledge. Following the methodological rules of unbiased observation in the laboratory, the dissection room and at the bedside, and of deducing conclusions from the facts collected, it can prove to deserve its recently granted external support. The problem with this scientific conception of medicine was that when data are accumulating and new facts multiplying, it became difficult to detect any cohesion and common perspective in the rapidly growing body of knowledge. It was also argued that the patient as the object of medicine is no longer adequately addressed since the conceptual tools of medicine are insufficient and too simple (Zimmerman, 1934).

One of the strategies to address these problems was a renewed self-interpretation of medicine as an art. Richard Koch for example defines medicine as *Heilkunst*, a healing art (Koch, 1930). Special attention must therefore be given to the personality of the doctor. The personal qualities of the artist determine his achievements. When the personal qualities of the physician are so important for the accomplishments of medicine, serious doubts could arise about the central role of epistemological issues in philosophizing.

III. THE ANTHROPOLOGICAL TRADITION IN PHILOSOPHY OF MEDICINE

The interpretation of medicine as an art evolved into a new conception of medicine as anthropological medicine. It was popular, particularly in Germany and the Netherlands, from approximately 1920 until 1960. Major representatives are F. Buytendijk, V. von Weizsäcker, V. von Gebsattel, H. Plügge, and P. Christian. Many advocates of anthropologically oriented medicine are practicing physicians and prolific writers with a broad interest in the humanities. Their main objective is to redefine and re-interpret medicine as a science of man. In the philosophical analysis of medical activities they applied and elaborated ideas and notions from several contemporary philosophical schools, particularly phenomenology (Husserl, Merleau-Ponty), existentialism (Marcel, Sartre), and philosophical anthropology (Scheler, Gehlen, Plessner, Strauss). Being interested in the human condition, and in particular its vulnerability and finiteness, they try to reflect upon human existence in its concrete specificity and ambiguity. Like Spicker (1976), they do not pretend to propose or discover the ideal image of the human being. Nonetheless, they attempt to identify what is anthropologically characteristic and common to human beings. But at the same time they are very much aware that any image is too abstract and clean, because in everyday reality the specific individual is always changing, pluriform, and not fully described by the designed image of man (Plügge, 1967). Given this theoretical point of view, advocates of anthropological medicine have not presented a clear-cut theory that is defended and elaborated by all representatives. Furthermore, as clinicians they prefer casuistry over systematic exposition, empirical practice over consistent theory, experience over grand theory. Their ideal is to change medicine from within, not from an external basis of general theoretical views. Individualized medical activities are prior to any systematic and theoretical approach. The publications of Viktor von Weizsäcker illustrate this; his subsequent books address practical problems which are encountered in clinical work and invite fundamental reflection (von Weizsäcker, 1941), they present patients' stories and cases (von Weizsäcker, 1947), and they focus explicitly on the ill person (von Weizsäcker, 1951); even the composition of the latter book (casuistry first, then theoretical exposition) illustrate this 'method.' Notwithstanding the lack of systematic unity in the works of advocates of anthropological medicine, there is a certain 'family of ideas' that became influential in the above-mentioned period. Recently, many studies have examined the basic tenets of this movement, and have argued for the relevancy of the ideas for contemporary medicine (Dekkers, 1985; 1994; ten Have and van der Arend, 1985; ten Have, 1983; 1987, 1994; Kasanmoentalib, 1989; von Uexküll, 1994; Verwey, 1990; Welie, 1994).

One basic idea in this anthropological tradition is that human beings cannot be subdivided in a physical and mental compartment. This is the rejection of the Cartesian notion of "body," advocated also by Spicker as the theoretical starting point for philosophy of medicine (Spicker, 1976). Although medicine has profited from this subdivision, it also has restricted itself to the human body, studying and explaining the body's physico-chemical machinery. This medical approach of the human body should be criticized because it reduces the human being to a specific animal species, and the human characteristics of the human body to its physical level of being. Even psychosomatic medicine has not overcome this anthropological dualism since any assumption of an interaction between body and mind implies a separateness of two substances. Dualistic thinking not only involves a reduced image of the human person, but it also has a more general tendency. Focus of critique is the general distinction made between object and subject. Von Weizsäcker, for example, strongly rejects the idea that there is an objective, real world, independent from an isolated, individual subject. A human being cannot relate him/herself to the world as an neutral observer. We cannot know the world in which we live, without at the same time changing it (von Weizsäcker, 1948).

Anthropologically oriented physicians argue that the methodology of the natural sciences is not fully appropriate in the context of health care. Medicine is the science of the human person. Scientific methods used to be abstract, analytical; they proceed from a model of linear causality. These methods also focus on intervention, control and manipulation, introducing the technical point of view of the engineer into the domain of disease and suffering. The conclusion is not that the causal thinking and technical approach of the natural sciences should not be allowed in medicine; on the contrary, they are highly valuable and useful; but medical thinking and practicing should not restrict itself to these scientific methods. In the words of Von Weizsäcker: "Medicine is not technology; it is technology, too" (von Weizsäcker, 1957, p. 32). The problem is that scientific methods cannot grasp what is essential to human beings. As a living organism, every person constitutes a whole, a meaningful entity, which is disconnected and disintegrated in abstract, analytical approaches. Buytendijk, for example, argues that from the mechanical viewpoint of the scientific method only those aspects can be explained of living organisms, that are not intrinsic to life itself (Buytendijk, 1961). This method can be applied on human beings, but it will naturally remain within a theoretical intermediate sphere, since it is always disengaging itself from human existence. Anatomical, physiological and biochemical research only determines the conditions, teaches us what is possible and probable, not what really happens. To examine the living being, we should participate in life, and focus on the purposeful coherence and interrelation, the significance of experience and conduct. If medicine wants to evolve into a science of the human person, it should overcome the usual distinction between the objective and subjective. That means that the subject should be introduced into medicine. As Buytendijk summarizes: If medicine is not objective, it is impossible; if medicine is only an objective science, it is inhuman (Buytendijk, 1959).

What is necessary, therefore, from a truly anthropological perspective, is a new, comprehensive understanding of disease. The current conception of disease in modern medicine is incomplete, according to anthropologically oriented physicians. The reason is not that the science of pathology is insufficient and not fully developed, but that pathology operates with inadequate notions and assumptions. Focusing on the causal mechanism of disease, pathology-based medicine cannot fully understand the ill person, because explaining disease also refers to the problem of the significance of a symptom, the meaning of a particular complaint. Science based medicine in fact hinders the insight that disease has meaning; in its approach the only relevant question concerns the pathogenesis and pathophysiology of the disease; the anthropological question 'Why here and now?' is irrelevant. The argument is not that medicine should more address the psychological dimensions of disease. Since dualistic thinking should be rejected, we cannot say that the body and/or the mind are involved in the disease process; it is impossible to find out 'where or who started the disease,' thereby recognizing the primacy of body or mind.

Anthropologically-oriented physicians argue that it is impossible not to think that disease has meaning. The body gives expression to the human person. Disease therefore gives voice to a threatened existence. Being ill primarily is an existential category; only secondarily, we can make any differentiation of organic and psychic suffering. Being ill is a way of being a human person. When we do not only have our life, but also give expression to it, when we do not merely have our bodies, but also are our bodies at the same time, then it is also the case that we are not only having our disease and suffering it, but also make our disease.

IV. THE RECENT PHASE OF BIOETHICS

The distinction between various phases within the evolution of modern philosophy of medicine not only brings some structure in the history of philosophy of medicine but it also shows the continuity of its basic preoccupations. The few historical studies available do not demonstrate any evolution of the discipline (Szumowski, 1949). Retrospectively, there seems to be a haphazard accumulation of issues addressed and views criticized. From today's perspective, it might even appear that philosophy of medicine is a very recent affair and that the present preoccupation with medical ethics shows a marked discontinuity with earlier efforts to philosophize about medicine.

The present-day domination of medical ethics is, however, continuous with the various traditions of philosophy of medicine since the nineteenth century in that it is moved by the same commitments and fundamental problems. This continuity will only be visible through examining the other traditions, in particular that of anthropologically oriented medicine which flourished prior to the current interest in ethics.

Although the anthropological orientation of philosophy of medicine has apparently been superseded by the growing interest in medical ethics since the 1960s, there is a marked continuity between both phases in philosophy of medicine. By concentrating on the subjectivity of the patient, anthropological medicine paved the way for the subsequent interest in ethical issues in health care. The programmatical demand of Von Weizsäcker to introduce the subject into the life sciences and medicine implied not only acknowledging the subjectivity of the knowing and acting subject (the physician) but also that of the object (the patient). Doctor as well as patient experience themselves as psychosomatic organisms. That means for example that medicine should examine the relation between personal biography and illness. The identity of medicine, therefore, is constituted through internal determinants, *viz.*, the personal qualities of its practitioners, but, more significantly, also through external determinants, *viz.* the individual qualities of patients. Medicine is a unique profession in systematically and methodically attending to the patient as an irreducible person.

The tradition of anthropological medicine made visible and uncovered the moral dimension of medicine. It did so by criticizing the presuppositions of the dominant conception of medicine as natural science, and by incorporating medical science's analytical methods and mechanistic image of man in a broader framework of an authentic science of man. The point is that anthropological medicine brought to the fore that the internal characteristics of medicine are not identifiable without involving the external context. In fact, it appears that anthropological medicine is itself a normative program, an extended argument for a normative science of life. This is evident for example in the Weizsäckerian concept of subject which includes "organic" and "pathic" subjectivity: the subject is an acting subject, performing biological acts, as well as a sentient and perceiving subject. The concept of pathic subject originates from the practical context of medicine in which life is not an event but an experience, not activity but passivity. The pathic subject has to suffer and endure the existence (it is *subjectum*). Von Weizsäcker's philosophy is an attempt to reconcile organic life (*zooè*) and biographical life (*bios*). Scientific medicine must recognize and do justice to the dual nature of the subject. Medicine's understanding of man is limited as long as it is concentrating upon the physiological, organic and ontic aspects of life. It has to accept that any human life ". . . fits within the 'pathic' categories of want, must, can, may and shall" (Kasanmoentalib, 1989, p. 328). These categories are in the end normative. They embody values and ideals that must be realized in human life to call it really human (*bios*). It is through the uncovering and construction of a normative dimension of medicine, as really essential to its identity, that the anthropological phase is in formal respect continuous with the ethical phase. Without a conception of medicine as an anthropological, and eventually normative science, the later upsurge of interest in moral issues would not have gone so fast.

So far with regard to continuity. It is also obvious that a considerable discontinuity exists between both phases. This discontinuity is manifest when we look at the material components of the normative dimension of medicine. Within the anthropological approach, the intended normative science of life is primarily dependent on the person of the physician. It is he who uses the criteria to appraise what subjectivity really means. As Von Weizsäcker's clinical case analysis shows, it happens many times that the meaning of illness is concealed from the patient himself, while the physician knows how to uncover its meaning from the patient's biography (von Weizsäcker, 1951). In practice, anthropological medicine has more or less a tendency towards paternalism. The patient is indeed introduced into the self-interpretation of medicine but his role is staged by the anthropological director.

Another point is the emphasis upon pathic subjectivity. Its connotation of passivity was more and more out of tune with the revolutionary development of medicine, particularly since the 1960s. In an era in which medicine has become an exemplar of technoscience and technical power, reflection should focus on notions as activity, intervention, control, and manipulation. The concept of pathic subject therefore opened up the normative dimension of medicine, but having done so, it was no longer productive in making sense of the normativity.

V. A RE-EVALUATION OF BIOETHICS

It is a commonplace that interest in the moral problems of medicine and health care has burgeoned over the last three decades, in the U.S. as well as in the majority of European countries. Although medical practice has always been the subject of ethical critique, two recent developments in particular have served to transform the traditional notion of medical ethics: 1) advances in medicine and health care due to the influence of biotechnology and technology-oriented medicine, and 2) the rapidly changing socio-cultural context marked by the prevalence of a plurality of values in western countries, especially those values that bear on the provision of health care.

Traditionally, 'medical ethics' referred to the deontology of the medical profession, to codes of conduct which consist partly of ordinary moral rules, partly of rules of etiquette, and partly of rules of professional conduct (Downie, 1974). In this sense bioethics has the following characteristics:

(a) It is essentially a set of problems that focus on the internal morality of medicine, *viz.*, those values, norms, and rules intrinsic to the actual practice of health care. Medicine is not considered a merely technical enterprise that can be morally evaluated from some exogenous standpoint. On the contrary, the professional practice of medicine always presumes and implies a moral perspective or point of view; therefore, what is judged to be sound medical practice is determined by the shared rules and standard procedures of the practice.

(b) It is related to the professionalization of medicine. Through this historical process emphasis is placed on the common good, and this was combined with an appeal to the self-interest of the members of the profession. Social recognition could only be gained on the basis of a strong internal organization and self-imposed standards of behavior. Self-regulation by physicians and a special style of life, structured in terms of high ideals, duties, and virtues, could promote the physician's image, and thus the power and prestige of each member of the medical profession.

(c) Since it was primarily concerned with explicating norms and formulating standards of professional conduct, medical ethics and etiquette have been segregated for a long time from general intellectual history (Fox, 1979). Moreover, before the 1960s, medical ethics was not a subject frequently discussed in public fora and the extant literature. Apparently, there was a consensus of opinion concerning the moral commitments of those who provided medical care, and the explication or codification of these commitments was regarded as the principal concern of medical professionals.

Since the 1960s, professional medical ethics has gradually detached from its traditional deontology (ten Have and van der Arend, 1985), although there are important phase differences between the U.S. and European countries; especially in Southern European countries the emphasis in medical ethics still is on "medical deontology." In Northwestern European countries and the U.S. professional medical ethics more and more is subsumed under 'health care ethics', or 'bioethics.' These new terms tend to indicate that the discipline of ethics not only includes problems that arise in the physician-patient relationship, but also a number of moral problems posed by other health care professionals, as well as moral issues created by the health care system, and public policy issues engendered by biomedical advances and the results of research.

The outcome of the above-mentioned transformation process is more salient in the U.S. than in most European countries. Moral problems in health care are generally approached in U.S. bioethics in a specific way, more analytic and applied than is usually the case in many Continental approaches. Today there is growing concern that the results of the aforesaid transformation are unsatisfactory. Professionalization and institutionalization of medical ethics received an enormous stimulus because both the adequacy and the relevance of medicine's internal morality were put into question. Professional ethicists have placed more and more emphasis on the crucial role of external morality: the principles, norms, and rules operative in society that bear on medicine and are frequently codified in law. Thus, for some, medicine and health care are nothing more than interesting "intellectual" phenomena with respect to which general ethical theories, principles, and rules may be applied.

This shift from internal professional to external morality and the predominant interpretation of medical ethics as "applied ethics" encouraged physicians to criticize present-day medical ethics for its lack of attention to the practical vicissitudes of health care, for its theoretical biases, and its conceptual alienation from clinical reality (Editorial, 1985;

Vandenbrouke, 1990). It is also claimed that the conceptual ground of medical ethics is too limited and even reductive when seen from the perspective of the tradition of philosophical ethics itself (Baier, 1985; Brody, 1988; Kass, 1990).

Moral theories and principles are necessarily abstract and therefore not relevant to the particular circumstances of actual cases, the concrete reality of clinical work, the experience of having and being a body, and the specific responsibilities of health care professionals. By appealing to principles, norms or rules, applied ethics may fail to realize the importance of concretely lived experience of health care professionals, as well as patients. The moral agent is taken to have an abstract existence. This point is critically elaborated by contemporary philosophers. Ethics, according to Bernard Williams (1988), does not respect the concrete moral subject with his personal integrity. It requires that the subject gives up his personal point of view and exchanges it for an universal and impartial point of view. This is, Williams argues, an absurd requirement, because the moral subject is requested to give up what is constitutive for his or her personal identity and integrity. The idea that knowledge of normative theories and principles can be applied to medical practice simply ignores the fact that moral concerns tend to emerge from experiences in medical settings themselves. A similar issue is raised by Charles Taylor in his recent work, in which morality and identity are considered two sides of the same coin (Taylor, 1989). To know who we are is to know to which moral sources we should appeal. The community, the particular social group to which we belong, is usually at the center of our moral experience. Even the use of ethical language depends on a shared form of life. The Wittgensteinian notion that our understanding of language is a matter of picking up practices and being inducted into a particular form of life is germane here.

In short, the criticism is that bioethics should become more appreciative of the actual experiences of practitioners and more attentive to the context in which physicians, nurses, patients, and others experience their moral lives, e.g., the roles they play, the relationships in which they participate, the expectations they have, and the values they cherish (Zaner, 1988). The physician-patient relationship is neither a-historical, a-cultural nor an abstract rational notion; persons are always persons-in-relation, are always members of communities, are immersed in a tradition, and participants in a particular culture. In most ethical discourses, there also is no recognition of the special experiences of embodiment; it seems as if the autonomous subject is without body; the body merely is the instrument through which the subject is interacting with the world; the subject is in full control of his or her body; the integrity of the body should therefore be respected because it is the prime vehicle of the autonomous person.

Given the criticisms noted above, an alternative approach to and conception of bioethics is clearly needed – a conception that provides a more comprehensive understanding of the nature, scope, method, and application of ethics in the contemporary health care context. It will be necessary to re-connect bioethics with both a general philosophical standpoint and the concrete practice of medicine. A renewed reflection on the significance of philosophical anthropology may be a road towards such reorientation of bioethics. A broader philosophical framework for a practicable bioethics is needed.

VI. RE-INTRODUCING PHILOSOPHICAL ANTHROPOLOGY

Alternative approaches in bioethics, such as narrative ethics, feminist ethics, hermeneutical ethics, care ethics, seem to have a common denominator: they focus on the connection between the internal and external morality of medicine, without reducing one set of norms and values to another. It is heuristically assumed that on the one hand there are specific values, norms and rules intrinsic to the actual practice of medical care (the "internal morality"), on the other hand, values, norms and rules prevailing in social, cultural and religious traditions that function as external determinants of medicine (the "external morality"). The dominant conception of bioethics proceeds from a too strong distinction between these two sets of values, norms and rules, as well as an over-estimation of the relevancy and importance of the external morality. In order to obtain a better understanding of the interaction of both moralities, it is necessary to establish a theoretical framework relevant to medical practice in order adequately to take account of the norms and values inherent in the practice of medicine, but it requires at the same time sufficient detachment in order to provide a critical normative perspective on medical practice.

The question is how this task can be accomplished. How to develop a theoretical perspective on medical ethical issues that connects philosophical reflection with the everyday realities of medical practice? Such perspective not only aims at elucidating specific bioethical problems, but it intends critically to examine various conceptions of bioethics that purport to deal with such problems. It should also make clear why and how such bioethical problems appear, reappear, and even disappear in medical discourse; why certain problems emerge in various health care practices and other do not; and how such problems can be discussed and even resolved during daily interactions between physicians, nurses, patients, hospital administrators, and others. I have pointed out elsewhere what steps can be taken to pursue the objective of reconnecting the internal and external moralities of medicine and healthcare (ten Have, 1994).

The first step is to examine the methods of clinical ethics in different health care contexts in order to obtain a better understanding of the *internal morality* of these practices. This will require to formally undertake both empirical research and philosophical investigations of concrete clinical settings (ten Have and Lelie, 1997).

The second step is to analyze and interpret the *external morality* governing health care practices. Making use of the results of recent social research and specific empirical investigations, this step requires the study of values, norms, and attitudes concerning medical-ethical issues.

The third step is the creation of *new theoretical perspectives* on health care practices. History of medicine as well as philosophy of medicine share a growing interest in de empirical realities of medicine. The so-called empirical shift in philosophy of science, some decades ago, has led to new approaches, e.g., several kinds of social constructivism (ten Have and Spicker, 1990). Ethics, philosophy and history of medicine may thus find common ground in creating new theoretical perspectives on health care practices. In any practice a complex set of activities guided by shared rules, cognition, action and normativity are inextricably linked. Focusing on the notion of *practice* as the common theoretical starting point, the interdependence of the disciplines as well as the specificity of their expertise will become apparent. Philosophy of medicine analyzes the cognitive components of health care practices: concepts, methods, and ideas. Medical ethics examines the activities and action-guides embodied in health care practices as well as the values embedded in such practices. History of medicine studies the construction and transformation of practices.

The fourth step is to develop a new conception of bioethics that illuminates and clarifies the complex interaction between the internal and external morality of health care practices. Here the explicit introduction of a philosophical anthropology will be necessary. As a particular domain of philosophy, ethics proceeds from empirical knowledge, *viz.*, moral experience. This experience is the embodied experience of the human being. The moral dimension of the world is first and foremost experienced. Moral experience is humanity's way of understanding itself in moral terms (van Tongeren, 1988). Ethics is therefore the interpretation and explanation of this primordial understanding. Before acting morally we must already know, at least to some extent, what is morally desirable or right. Otherwise, we would not recognize what is appealing in a moral sense. On the other hand, what we recognize in our experience is typically unclear and in need of further elucidation and interpretation. In short, we all approach the moral dimension of the world from a set of prior understandings; they form the basis of our interest in what at first seems odd and strange to us, requiring us to continuously reconstruct the moral meaning of our lives.

VII. CONCLUSION

The overwhelming impact of moral issues in post-modern health care as well as the urgency to address and resolve such issues, might induce a certain amnesia concerning the various perspectives that used to be characteristic of philosophy of medicine. In order to understand and clarify the problems of present-day medicine, epistemological, anthropological and ethical perspectives all are indispensable. The identity of medicine nowadays is no longer solely based upon internal characteristics and intrinsic norms and values, but also upon external determinants and values. The unique object of medicine is the human embodied being, a person *having* as well as *being* a body. Fundamental anthropological phenomena (such as mortality, invulnerability, and infirmity) are specifically articulated in contemporary medicine. Philosophy of medicine presupposes that these phenomena have been explored philosophically, as Stuart Spicker has emphasized on many occasions. The question what medicine essentially is, is no longer answered by medicine itself but requires more general anthropological exploration. This requirement is especially urgent for the further development of bioethics. At present, medicine is identified as a complex of meaningful knowledge and practices, leading to significant value problems and normative questions. But these questions presuppose one common denominator: the human being him/herself. Philosophical elucidation of what it is to be a human being and to be treated as a human being, will therefore be inevitable for the further development of philosophy of medicine and bioethics in the next millennium.

Department of Ethics, Philosophy and History of Medicine
Faculty of Medical Sciences
Katholieke Universiteit Nijmegen
Nijmegen, The Netherlands

BIBLIOGRAPHY

Baier, A.: 1985, *Postures of the Mind. Essays on Mind and Morals*, Methuen, London.

Brody, B. (ed.): 1988, *Moral Theory and Moral Judgments in Medical Ethics*, Kluwer Academic Publishers, Dordrecht, the Netherlands.

Buytendijk, F. J. J.: 1961, *Academische redevoeringen*, Dekker & Van de Vegt, Utrecht/Nijmegen.

Buytendijk, F. J. J.: 1959, 'De relatie arts-patiënt', *Nederlands Tijdschrift voor Geneeskunde* 103, 2504–2508.

Dekkers, W. J. M.: 1985, *Het bezielde lichaam. Het ontwerp van een antropologische fysiologie en geneeskunde volgens F. J. J. Buytendijk*, Kerckebosch, Zeist, the Netherlands.

Dekkers, W. J. M.: 1994, 'F. J. J. Buytendijk's Concept of an Anthropological Physiology', *Theoretical Medicine* 16(1), 15–39.

Downie, R. S.: 1974, *Roles and Values. An Introduction to Social Ethics*, Methuen, London.

Editorial: 1990, 'Medical Ethics: Should Medicine Turns the Other Cheek?', *The Lancet*, Oct. 6, 846–847.

Fox, D. M.: 1979, 'The Segregation of Medical Ethics: A Problem in Modern Intellectual History', *Journal of Medicine and Philosophy* 4, 81–97.

ten Have, H. A. M. J.: 1980, 'Wijsbegeerte der Geneeskunde', *Algemeen Nederlands Tijdschrift voor Wijsbegeerte* 72(4), 242–263.

ten Have, H. A. M. J.: 1983, *Geneeskunde en filosofie. De invloed van Jeremy Bentham op het medisch denken en handelen*, Uitgeversmaatschappij De Tijdstroom, Lochem, the Netherlands.

ten Have, H. A. M. J.: 1983, 'Antropologische Geneeskunde. Silhouet van een Specifiek Menselijke Geneeskunde', *Metamedica* 62, 10–21.

ten Have, H. A. M. J. and van der Arend, A.: 1985, 'Philosophy of Medicine in the Netherlands', *Theoretical Medicine* 6, 1–42.

ten Have, H. A. M. J.: 1987, 'Afscheid van antropologische geneeskunde?', *Metamedica* 66, 245–259.

ten Have, H. A. M. J. and Spicker, S. F.: 1990, 'Introduction', in H. A. M. J. ten Have, G. K. Kimsma, and S. F. Spicker (eds.), *The Growth of Medical Knowledge*, Kluwer Academic Publishers, Dordrecht, pp. 1–11.

ten Have, H. A. M. J.: 1994, 'The Anthropological Tradition in the Philosophy of Medicine', *Theoretical Medicine* 16(1), 3–14.
ten Have, H. A. M. J.: 1994, 'Principlism: A Western European Appraisal', in E. R. DuBose, R. P. Hamel and L. J. O'Connell (eds.), *A Matter of Principles? Ferment in U.S. Bioethics*, Trinity Press International, Valley Forge, Pennsylvania, pp. 101–120.
ten Have, H. A. M. J.: 1997, 'From Synthesis and System to Morals and Procedure: The Development of Philosophy of Medicine', in R. A. Carson and C. R. Burns (eds.), *Philosophy of Medicine and Bioethics: A Twenty-Year Retrospective and Critical Appraisal*, Kluwer Academic Publishers, Dordrecht, The Netherlands.
ten Have, H. A. M. J. and Lelie, A.: 1998, 'Medical Ethics Research between Theory and Practice', *Theoretical Medicine and Bioethics* 19, 263–276.
Kasanmoentalib, S.: 1989, *De dans van dood en leven. De Gestaltkreis van Viktor von Weizsäcker in zijn wetenschapshistorische en filosofische context*, Kerckebosch, Zeist, the Netherlands.
Kass, L. R.: 1990, 'Practicing Ethics: Where's the Action?', *Hastings Center Report* 20, 5–12.
Koch, R.: 1930, *Philosophische Grenzfragen der Medizin. Der Begriff der Medizin*, Vorträge des Instituts für Geschichte der Medizin an der Universität Leipzig, G. Thieme Verlag, Leipzig, vol. 3.
Plügge, H.: 1967, *Behagen en onbehagen. Bijdrage tot een medische antropologie*, Het Spectrum, Utrecht/Antwerpen.
Risse, G. B.: 1976, '"Philosophical" Medicine in Nineteenth Century Germany: An Episode in the Relations between Philosophy and Medicine', *Journal of Medicine and Philosophy* 1(1), 72–92.
Spicker, S. F.: 1975, 'The Lived-body as Catalytic Agent: Reaction at the Interface of Medicine and Philosophy', in H. T. Engelhardt and S. F. Spicker (eds.), *Evaluation and Explanation in the Biomedical Sciences*, D. Reidel Publishing Company, Dordrecht, the Netherlands, pp. 181–204.
Spicker, S. F.: 1976, 'Terra Firma and Infirma Species: From Medical Philosophical Anthropology to Philosophy of Medicine', *Journal of Medicine and Philosophy* 1(2), 104–135.
Spicker, S. F.: 1986, 'Cognitive and Conative Issues in Contemporary Philosophy of Medicine', *Journal of Medicine and Philosophy* 11, 107–117.
Spicker, S. F.: 1990, 'Invulnerability and Medicine's "Promise" of Immortality: Changing Images of the Human Body during the Growth of Medical Knowledge', in H. A. M. J. ten Have, G. K. Kimsma, and S. F. Spicker (eds.), *The Growth of Medical Knowledge*, Kluwer Academic Publishers, Dordrecht, the Netherlands, pp. 163–175.
Szumowski, W.: 1949, 'La philosophie de la médecine, son histoire, son essence, sa dénomination et sa définition', *Archives Internationales d'Histoire des Sciences* 2, 1097–1139.
Taylor, Ch.: 1989, *Sources of the Self. The Making of Modern Identity*, Cambridge University Press, Cambridge.
Van Tongeren, P.: 1988, 'Ethiek en Praktijk', *Filosofie & Praktijk* 9, 113–127.
von Uexküll, T.: 1994, 'Anthropology and the Theory of Medicine', *Theoretical Medicine* 16(1), 93–114.
Vandenbroucke, J. P.: 1990, 'Medische Ethiek en Gezondheidsrecht; Hinderpalen voor

de verdere toename van kennis in de geneeskunde?', *Nederlands Tijdschrift voor Geneeskunde* 134, 5–6.

Verwey, G.: 1990, 'Medicine, Anthropology and the Human Body', in H. A. M. J. ten Have, G. K. Kimsma, and S. F. Spicker (eds.), *The Growth of Medical Knowledge*, Kluwer Academic Publishers, Dordrecht, the Netherlands, pp. 133–162.

von Weizsäcker, V.: 1941, *Klinische Vorstellungen*, Hippokrates Verlag, Stuttgart.

von Weizsäcker, V.: 1947, *Fälle und Probleme*, F. Enke Verlag, Stuttgart.

von Weizsäcker, V.: 1948, 'Grundfragen Medizinischer Anthropologie', in V. von Weizsäcker: 1950, *Diesseits und Jenseits der Medizin*, K.F. Köhler, Stuttgart, pp. 136–166.

von Weizsäcker, V.: 1951, *Der kranke Mensch. Eine Einführung in die Medizinische Anthropologie*, K.F. Köhler, Stuttgart.

von Weizsäcker, V. and Wyss, D.: 1957, *Zwischen Medizin und Philosophie*, Vandenhoeck & Ruprecht, Göttingen.

Welie, J. V. M.: 1994, 'Viktor Emil von Gebsattel on the Doctor-Patient Relationship', *Theoretical Medicine* 16(1), 41–72.

Williams, B.: 1988, 'Consequentialism and Integrity', in S. Scheffler (ed.), *Consequentialism and Its Critics*, Clarendon Press, Oxford, pp. 20–50.

Zaner, R. M.: 1988, *Ethics and the Clinical Encounter*, Prentice-Hall, Englewood Cliffs, NJ.

Zimmerman, H.: 1934, 'Ueber Aerztliche Synthese', *Klinische Wochenschrift* 13, 453–456.

MARX W. WARTOFSKY

BODIES, BODY PARTS, AND BODY LANGUAGE: REFLECTIONS ON ONTOLOGY AND PERSONAL IDENTITY IN MEDICAL PRACTICE

I. INTRODUCTION

A favorite bit of patter (gleaned from a Yiddish comedian at a summer camp of my childhood) is the following opening line:

"Before I begin speaking, I would like to say a few words . . ."

The (preparatory?) few words here are intended to frame the questions posed in this essay, and to suggest what motivates them. I begin with the transparent fact that the body is the ubiquitous object of medical practice, and often, not the body as a whole, but its parts – an organ, a tissue, a particular function or system. Yet, as everybody knows, patients are not simply bodies but persons – individuals as distinctive and unique in their identity as they are generic in their species characteristics; hence, extraordinarily complex entities, from the point of view of what we may call the ontology of medical practice. The medical practitioner is expected to have a fairly sophisticated knowledge of what the relations are between bodies as wholes, and their parts, but is also faced with the need to have at least a tacit or working understanding of what the relations are between the bodily identity of a patient or subject, and that individual's personal identity. An older but still popular way of marking this distinction between the body and the person is the simple-minded division between medicine proper (which deals with the body proper – the real object of medical practice), and bedside manner (which deals with the person – a sort of complicated add-on). A current variant of this view is emblazoned on the sides of MTA public transportation in New York City, as an advertisement for a private health organization. It reads: "Medicine is a science – Caring is an art." A related and more modish distinction is that between "reductionism" (which "reduces" the subject of medical practice to a merely physiological or biochemical "object"), and "holism" (which takes into account the whole person, and the whole organism in its natural and social environment). Indeed, the critics of this sort of reductionism often characterize it as the "medicalization" of the patient or the subject, as if the term "medical"

H. T. Engelhardt, Jr. (ed.), The Philosophy of Medicine, 37–55.

already connoted this transformation of the patient from a "Thou" into an "It."

My own view is that it is by now a theoretically stale business to counterpose reductionist and holist concepts and practices in medicine. Everybody (or almost everybody) acknowledges the sins of reductionist abstraction, where the medical subject is construed or treated as a mere body, or worse yet, as a body part – an organ, a tissue, a limb – without regard for the patient as a "whole person." At the same time, most practitioners would acknowledge that effective treatment of disease or injury requires an abstractive focus on the disease entity or the insulted body, in terms of specific location and function; biomedical research surely owes many of its great advances to a relentless reductionism in its focus on the parts, and often on the smallest parts. Concern for the whole person won't set a broken leg or cure a viral infection or discover the causes of genetically transmitted diseases. The question, then, is not whether one should opt for reductionism rather than holism as the methodology (or the ontology) of choice, or as an interpretive matrix for diagnostic or therapeutic purposes. Rather, it is a question of understanding what the relations are between parts and wholes, and between bodies and the persons whose bodies they are. The motive is to sort out some of the complexities and puzzles that arise here, specifically as they bear on the practice of medicine, i.e., on the relations between practitioners and the objects of their practice.

In this context, I want to pose three related questions of a very general sort, but to limit my discussion here to their relevance for medicine, or more specifically, for an ontology of medical practice:

(1) What constitutes bodily identity? That is to say, what constitutes a given body – e.g., of a human being – as having an individual identity as that body and not some other?

(2) What constitutes the personal identity of an individual human being, and how, if at all, is that related to bodily identity?

(3) What constitutes the identity of a body-part, and what relation, if any, does that body-part identity have to the personal identity of the body of which it is a part?

I will make some simplifying assumptions in the discussion that follows. First, the bodies and body parts which concern us here are those of human beings, i.e., of distinct human individuals with personal identities (and all that goes along with that), insofar as they are objects (or subjects) of medical practices. (I am not talking about physical bodies in general, or even the bodies of organisms in general.) Second, I will assume a range of medical practices (in the plural) rather than some canonical or uniform practice with essential features characteristic of all of medicine. This is not to say that there are no shared features across the range of medical practices. It does presuppose, however, that there are different conceptions and operational definitions of bodies, body-parts and personal identity at work in medical practice. (Here I use the term in the singular as a generic term appropriately hedged by my anti-essentialist caveats.)

To set the scene, I begin (in section 2) with some preliminary reflections on cases in which concrete and practical questions of bodily and personal identity arise. Then (in section 3), I proceed to an exploration of conceptual problems concerning wholes and parts, and (in section 4) consider the questions concerning bodily and personal identity in this light. Finally (in section 5), I consider whether commitments to one or another of the alternative ontologies makes any difference in medical practice or not. In other words, do different "body-languages" reflect different modes of relation between practitioners and their patients, and if so, what are the consequences of such differences, for diagnosis, treatment and care?

II. BODIES, BODY PARTS AND PERSONAL IDENTITY

The reconstruction of personal identity from body parts is an aspect of forensic medicine. This is the often grisly business, much in the public eye recently, of identifying victims of war, or of violent crimes, or of airline crashes or other accidents, whose remains are mutilated beyond normal recognition, or are only partial, or conversely, it is the business of identifying perpetrators of a crime of violence (e.g., rape or murder) by tell-tale identifying clues: blood, semen, fingerprints, voiceprints, or other cellular or organic parts of the whole, where such parts may uniquely identify, or contribute to the identification of the whole person.

Another context in which the personal identity of body parts raises conceptual as well as medical and biological issues is that of organ transplants. In such cases, a foreign body part, belonging originally to another person (or in some cases to an animal) is surgically joined to a body as a replacement for an original body part which is defective, or has been destroyed, or is missing for some other reason (e.g., as a congenital defect). Here, the question of the "personal identity" of the body part is not simply a conceptual issue – ("Whose body part is it? Does it acquire a new 'identity' derived from the host body to which it has been transplanted?") – but a biochemical or physiological question of immunological acceptability by the host-organism. "Recognition," as a matter of personal identity, seems in this case to be not so much a matter of unique individual identity as a person, as it is what one might designate as *immunological identity*, i.e., as acceptability or recognizability as compatible, at the level of cell-recognition, or of antigens. Here, there is a margin of medical manipulation by means of the suppression of immune reactions to implanted or transplanted tissue, until some adaptation to the "foreign" invader can take place, and the transplanted body-part becomes a "naturalized citizen" of the organic community of the body in question. In this case, the "personal identity" of a body part is at the level of cellular or subcellular biochemical processes, and, to all appearances, far removed from the question of the personal identity of the body as a whole. And yet, what may without difficulty be characterized as the personal identity of a body part, in the cellular or biochemical senses just described, may be constitutive in a complicated way of the features of identity we reserve for the description of whole individuals or persons. Thus, for example, we may say that it is Smith's liver, or Smith's visual cortex, or his aorta, that (together with the rest of Smith's body parts) make up the person, Smith. Or we may say that the personal identity of the human individual, as a holistic characteristic, permeates or interpenetrates all the parts of the whole, or enough of them, to determine their identity with the whole even at the lowest levels of biochemical or immunological response.

Though immunological compatibility is more probable with genetically similar donors – e.g., twins or close kin – and is therefore not necessarily a matter of the personal identity of a unique individual, it still raises the more general question posed here: how far down the levels of biological organization does personal identity extend? For example, once the graft, the transplant, the implant, has "taken" as a body part, has it then acquired the "personal identity" of the whole of which it has become a part? Or does the implanted body part continue to remain, so to speak, a tolerated alien, a legal immigrant, but never a full-fledged ("native born"?) citizen of the body politic?

There is one additional case of a body part which appears to be different from the previous ones. This is, of course, the mechanical, or electronic, or electrochemical body part or prosthetic device, made of metal or plastic or some other material, which functions as a replacement or substitute for an original but defective or missing organ, but has itself no biological identity as tissue, or organ. Yet it is neutral or biochemically inert enough not to cause a rejective immune reaction. Here, the replacement mitral valves, heart chambers, pumps, conduits, gates, connective pins, wires, springs, artificial limbs, voice boxes, dentures – the whole armamentarium of prosthetic devices as body parts – all become functionally "identical" with the body parts they replace. But these helpful invaders seem clearly enough not to be candidates for the sort of personal identity of body parts which, I have suggested, characterizes organic transplants, grafts, etc.

One may ask: Why not? Isn't successful incorporation and functional equivalence as pragmatically and operationally sufficient a criterion of the alleged identity with the body as the immunological or the strongly personal identity we considered earlier in the organic cases? Or, contrariwise, does this case serve as a counterexample to the earlier claims for the "acquired" body-part identity of the medical replacements or substitutions for the original body parts? Don't all these body-part replacements, whether organic or not, remain just what they are: foreign elements which have been successfully incorporated into the host body, tolerated for the functions they perform, but never full citizens?[1]

The whole issue of body-part identity which I have addressed here may seem a totally empty one, for what difference could it possibly make, whether one regards the successful transplant or prosthetic device as having "personal identity" or else as nothing more than a functional addition whose "identity" is not even in question, as long as it works? Isn't this whole discussion based on a confusion between two separate and independent meanings of the term "personal?" One has to do with the identity of persons, the other with the biological or immunological identity of a given individual *body*, and thus is "personal" only in the sense of being uniquely identifiable, or functionally compatible, as in cell-recognition. That is to say, shouldn't personal identity be distinguished from bodily identity? In short, isn't there a confusion of levels of reference here? Further, isn't there also a confusion between two uses of the term "identity," one a strong sense (as in "self-identity" or "personal identity"), the other, a weak sense (as in "functionally a part of . . ." or "belonging to")?

These objections stem from a pragmatic consideration of what difference it would make in medical practice to adopt one view of the identity of body parts rather than another. On such an approach, if it made no difference at all, and remained no more than a choice among ways of speaking, or between one "body-language" and another, then ontologically, there would be no difference; or to put it differently, the extension of both terms of reference would be the same, though their intensions might differ. Thus, even if two physicians took themselves to mean different things in talking about the relation of a body part to the person it belongs to, or to whom it has been transplanted – one asserting that it has now become a proper constituent of the personal identity of that individual, while the other maintains that it remains an artificial part having no more of a relation to that person's identity than that person's watch or credit card, i.e., simply "belonging to" or "available for use by" that individual – if both physicians then did the same things in effecting the transplant, or in following up on its viability and its use, their alternative ways of conceiving of this body part identity would make no difference in their actual practice. But then, suppose these different conceptions did make a difference in practice? What difference could that be? I postpone a consideration of this possibility until later, but add the teasing remark that I will suggest that such differences in construals of identity do make a difference in medical practice, and one that doesn't simply reduce to differences in bedside manner.

In order to pursue this reflection on bodily and personal identity further, we need to consider the matter of the relation of wholes and parts, to which we now proceed.

III. WHOLES AND PARTS: WHAT'S AT ISSUE

There is no such thing as a "part-in-itself" or a "whole-in-itself," except for purposes of deliberate abstraction from the systemic relations of parts and wholes. The very concepts form an interdependent pair, like Descartes' well-known example of mountains and valleys, or Austin's, of a pair of pants. In short (and redundantly): to be a "part" is to be part of a whole, and to be a "whole" is to be made up of parts. (One may want to argue that there are seamless wholes that are irrefragible unities, like Parmenides' One, or the soul in Descartes' account. But though they are one and indivisible, they are not "wholes" in the neat, constructed version I have in mind here, so I stipulate that by "Whole" here I mean a constituted unity, made up of parts.)

Aristotle's classic formulation of the part-whole relation (in the *Politics*, where he is discussing the relation of the individual to the state, or *polis*), makes use of an anatomical/physiological analogy to support a strongly holistic view. He says,

> the state is by nature clearly prior to the family and to the individual, since the whole is of necessity prior to the part; for example, if the whole body be destroyed, there will be no foot or hand, except in an equivocal sense, as we might speak of a stone hand; for when destroyed, the hand will be no better than that. But things are defined by their working or power; and we ought not to say that they are the same when they no longer have their proper quality, but only that they have the same name (Aristotle, *Politics* I. 2, 1253a19–26, B. Jowett (trans.)).

In its modern versions, holism proceeds from the insight that diseases or injuries of a part – a bodily part, or a specific pathology as part of a living system, or a particular susceptibility or disposition to react to certain substances adversely – cannot be understood, or diagnosed or treated properly, unless they are understood in relation to the whole organism, or the whole person, or sometimes even in relation to the whole family or the whole social group, e.g., in the case of hereditary or of even wider genetically or occupationally defined groups (e.g., in Tay-Sachs disease or in sickle-cell anemia, or in silicosis). This holistic emphasis is reflected in the kind of medical history-taking which provides an essential feature of diagnostic procedures.

Examples abound of alternative emphases, however. The reductionist tradition, from its earliest Greek models, accounts for wholes as combinations of ontologically prior parts – characteristically, the "elements" or "atoms" out of whose coming together all things are formed. The successes of reductionism, in classical physics in the seventeenth and eighteenth centuries, in chemistry, by the nineteenth, and in biochemistry and genetics in the twentieth, would seem to argue for the concentration on more and more isolated and detailed study of parts of ever greater specificity as far down the scale of nature (or in the case of human subjects, as far down the scale of human constitution) as the phenomena to be studied allow. Genetics, while it may indeed be studied "holistically" as a statistical phenomenon of populations as ensembles, i.e., as "wholes," rests also on the reductive biochemical model of DNA, as a polymer entirely constructible in terms of its parts and their location. Similarly, while epidemiological studies focus on the statistical or causal features of diseases in populations, through time, they rest as well on the reductionist analysis of the biochemistry of pathogens, i.e., on the subcellular or even molecular structures and activities of bacterial and viral agents in a host system.

Clearly, "reductionism" and "holism" as strategies of research or matrices for interpretation, in the sciences generally, or in the biomedical sciences specifically, are not mutually exclusive choices. They are, instead, complementary, and sometimes one and sometimes the other is called for. Such a methodologically tolerant approach may be characterized as pragmatic: no deep ontological commitment about the nature of things depends on which choice one makes. Rather, it is a question of appropriateness to the task at hand. and more than this, in sophisticated hands, the interrelationship or the systematic interdependency of the two approaches allows for a richer and more complete comprehension of the subject of research, e.g., the human body as a structure of constituent parts, and at the same time, as the bearer of the personal identity of unique human individuals, as wholes.

Trouble begins when the choice among such alternative emphases or perspectives is seen, not only as a matter of methodological preference (based on adequacy of explanation, or on heuristic value in biomedical research, or on efficacy in diagnostic or therapeutic medical practices), but instead as a matter of what we may call ideological difference – i.e., as a question of ultimate and mutually exclusive metaphysical characterizations of the nature of the objects under consideration. Thus, for example, a radical physical reductionist might argue that "personal identity" is only an illusory fiction – like the "self" or the "mind" – foisted upon what in reality is "nothing but" a certain combination of cells, themselves reducible to more elementary physical parts, and furthermore, that what we call "personhood" is exhaustively accounted for by an (in principle) "complete" description of the human body as a biochemical complex, or as a physico-chemical entity. Alternatively, a robust holist might argue that the body and all the so-called "parts" of a human being are in themselves "nothing but" the particular expressions or externalizations or aspects of a totality whose "being" is in reality that of a personal identity, i.e., a unified and indivisible "self" (or "self-consciousness") which is a whole of which the body and its parts are no more than the external manifestations. The differences here are what we may characterize as ideological.

However, what is of concern here, in the relation between reductionist and holist approaches, is not this ideological debate about the ultimate metaphysical nature of living bodies or of persons. Rather, it is the difficulties that arise in the practice of medicine, in dealing with the various levels of organization of the living body, from molecule to mind, so to speak. The conceptual difficulties and complexities that arise in understanding the relation of whole to part seems to me to be less a matter of ideology, than of the necessary division of labor which is the condition for the intensive specialization that deep research and competent diagnostic and therapeutic practice require. Relations between holist and reductionist methodologies in these contexts is, at least in part, a matter of the strategy and tactics of research and treatment, accommodation to the demands of the subject matter, and the cooperation of researchers, at different levels of the organism (from cell to whole body to population), in reconstructing the "big picture." These alternatives, then, are not simply abstract philosophical or methodological ideas, but products of a necessary division of labor in medicine. But this division brings with it, in addition to the benefits of deep specialization, the risks of one-sidedness, of fragmentation of the medical subject into parts treated as abstractions in practice, without relation to the whole. To overcome this, two things are required: first, a systematic mode of cooperation among the specialist-practitioners in which they reconstruct the whole collaboratively; second, the generalist, as a kind of "specialist" whose specialty is precisely an ability to relate the parts and the whole, and to make clinical judgments on the basis of this knowledge as to what treatment of the parts is necessary to the well-being of the whole. In the optimal practice of medicine, these two requirements have long been institutionalized, and the norms of good health care demand the provision of just these features. But what underlies such practical norms theoretically, is some clear conception of the identity of the subject of these various practices. I have suggested that this is a complex identity, and that in order to understand it, one needs to understand the constituent identities of the patient: as an embodied person some of whose parts are the proximal objects of treatment.

IV. BODILY IDENTITY AND PERSONAL IDENTITY

A remarkable feature of the human individual – i.e., of the individual human organism, whether as a living body, or even in death – is its utter distinctiveness or unique individuality. We tend to characterize this unique individuality as "personhood" or "personality," or to mark it as "personal identity," and we just as often assimilate this to the bodily identity of the person, as if personal identity and bodily identity were coextensive. In a certain sense this is true. Every person has a body uniquely identified as that person's body, and every human body is the body of a person. (The relation between these two entities then can be expressed or represented logically as a biconditional: ($x \supset y$ & $y \supset x$).) But that is not yet to say that persons and bodies are identical. It would mean, however, that if we could establish the identity of a body, we would then have determined it to be the (exclusive and unique) body of a particular person, and not that of any other person. Similarly, if we were to identify a certain person, we would at the same time have established the identity of that person's body, as the one that person is "in," or which "belongs to" that person, or which "is" that person (as the body "identified with" that person, where the locution "identify with" does not mean the same as "identical to" or even "identical with" – a nice distinction in idiomatic English). Of course, one may take the view that "person" and "body" are two names for the same thing, that, in fact, there is extensional identity of the two, and that the distinction is "only" intensional, a conceptual distinction "in the mind," so to speak, or in the different ways in which the terms "person" and "body" are used to refer to the same (identical) entity, either as manner of speaking, or as addressing distinguishable aspects or properties of the same thing. These distinctions take us perilously close to more general questions about identity ("mind/body," "Evening Star/Morning Star," etc.) that have been of concern to philosophers. While these considerations are broadly relevant to our questions here, we bracket them, or interpret them narrowly for our more specific purposes.

In an earlier section, we reflected on the applied contexts in which the questions of personal identity or bodily identity arise, with respect to body parts. Here, we return to these questions again, but more generally, as they were posed at the outset: what constitutes bodily identity? What constitutes personal identity, and how is it related to bodily identity? And what, if anything, constitutes the bodily or personal identity of body parts? I approach these questions obliquely, from the point of view of practices of identity, namely, how do we go about determining such identities in applied cases? The strategy here is not to offer some definitions or theories about identity in these various modes, but to generate the theoretical characterization from what is presupposed or implicated in our practices of establishing identity.

We begin with bodily identity. The normal or commonsensical presupposition here is that each body, for all the generic or species characteristics its shares with all other human bodies, is an individual and unique body. This entails that it has a life-history or biography, as a self-identical entity, that this history is ontologically continuous in time and space – i.e., that there are no gaps in that body's existence during its life process, from birth to death, and that it retains some mode of bodily identity even as a cadaver, or as an embryo. The operational identification of an individual body then proceeds by means of knowledge about its existence. This includes the external evidences of its identity e.g., in terms of physical or biological traits, its location at given times, features of that body's medical history (blood type, scars or other bodily traces of past injuries or diseases or treatment, X-ray or other records, etc.) all of which can be matched with observation of the body in question. There is of course no limit to the sophistication of biochemical methods which may be used to determine bodily identity in these terms, as our science and biomedical technology progresses. Then, there is, in addition to what may be called the biomedical recognition of the body's identity, the physiognomic recognition of the body in terms of such observable features as facial appearance and facial expression, speech traits, posture, gait – all the evidences of the police "line-up" or of family identification of kin, where recognition depends on previous physiognomic familiarity, that subtle mode of fallible knowledge that often eludes our analysis, but remains a profound element of our cognitive powers. There is still another mode of identity which takes us close to the borderline between bodily and personal identity, which concerns what we may dub cultural recognition, as a corollary to physiognomic recognition. This concerns the identification of a body in terms of such things as manner, or style, or distinctive modes of conduct, or bodily behavior. But this introduces another question. Is there a dividing line between the body as a biological organism and the person as some sort of psycho-social or cultural construction?

Bodily traits need not be innate or genetic. They may be acquired, as one's medical history shows, in terms of the experiences of illness or injury, etc. And, in fact, bodily traits, such as posture, gait, speech or voice characteristics, physiognomy (including such features as facial expression, look or gaze, and *embouchure*), all become subjects for anthropological investigation as the bodily aspects of cultural difference. Thus, biology passes over into culture even at the level of bodily identity. And if mind or cognition is itself construed as a bodily trait, e.g., as "nothing but" an activity or a property of the brain (as it may well be, by reductionist materialists or physicalists, or by mind-body identity theorists), then the psycho-social features of personality, or traits of consciousness in general may be conceived as constituting part of the body's identity. Of course, in this latter case, bodily and personal identity would be conflated, and all the presumably distinctive characteristics of personal identity would be exhaustively taken up (or taken down?) into features of bodily identity, *tout court*.

By contrast, personal identity seems to be characteristically related to self-identity, not simply in the reflexive sense of "self" (where we would take it as a truism that everything is identical with itself) but in the sense of self-awareness of one's identity, or as involving at least some rudimentary mode of self-consciousness. In practice, we go about determining someone's personal identity not simply by asking others questions about given individuals, but by inquiring directly of them who they are, or by gathering other personal information from them that would help indirectly to establish their identity. Now it would seem that this requirement of self-identification is a very strong one, obviously going beyond bodily identity, which may be ascribed to an individual simply in virtue of bodily traits of the biomedical or even physiognomic sort considered earlier, and provided entirely by observation or examination by others, hence requiring no first-person reports or information. However, as we have seen, bodily identity in terms of cultural traits begins to suggest features of an individual which would be hard to conceive without attributing consciousness or perhaps even self-consciousness to such an individual. What is at issue, of course, is what, if anything, distinguishes the personal identity of a human being from his or her bodily identity. And perhaps here we can do with something less than self-consciousness. We may proceed piecemeal, or constructively, to see what we may need to add to the bodily identity of an individual in order to ascribe personal identity to that individual as well. But note that if we ask the question in this way, we have already tacitly accepted one theory of personal identity, and perhaps the most commonsensical one: namely, that personal identity cannot be the feature of a disembodied "I," but rather presupposes a body which is the bearer of this identity, if that body has the appropriate criterial properties. But what are these?

In this bottom-up reconstruction, at the "lowest" limit, so to speak, we may want to say that the minimal criterion of personhood or personal identity is genetic identity, in the following sense: if a human being exists at all, then he or she was born from parents (however arranged, by natural or artificial insemination, borne by a mother or by a surrogate, etc. etc.), and as such, the offspring literally inherits a personal identity as the offspring of these parents (or of this sperm and egg, whatever their lineage!). No other requirement for personal identity is needed beyond this genetic determination as a unique human being. Alive or dead makes no difference, then, either. Personal identity, at this lowest limit, entails no more than a biologically objective circumstance, and there is no requirement for "personality" or "consciousness" or "selfhood," and certainly not for self-consciousness. What makes this "personal" rather than merely "bodily" is the fact of human parenthood, where personhood is simply passed along from one's progenitors, along with the DNA.

This is surely a problematic view, since it would seem to confer personal identity on stillborn infants, or acephalic or decorticate neonates, or cadavers. But perhaps this is right, since, in the case of victims of crime or accident, it is not simply the identity of the body that is in question, but "whose body it is," as we would tend to say, and this seems to suggest strongly that we are concerned to know who the person is that this body belongs (or belonged) to, and what its relation is to others, e.g., as next of kin, etc. Similarly, with stillborns and the like, the network of family or other social or legal relations, including here the relations to any medical personnel involved, to government agencies, to insurance companies, all establish a complex identity which is not simply the biological identity of a body, nor even the genetic identity conferred by the kinship relation of parenthood. Perhaps, then, personal identity requires no more than this culturally or socially or legally constructed set of relations, based on parental descent, but going beyond it. Just as the corporation is treated as a "person" in contract law, as a legal fiction, so too the human body is treated as a "person" in human society, as a social fiction.

Yet, this minimal account, based on extrinsic "objective" criteria – genetic or sociocultural – seems to miss a crucial feature of personal identity, namely, the awareness of oneself as a human being, i.e., as having a self-conscious identity as a person. The classic analysis of this self-conscious species identity and of the process of its development, not only as an "I" or a self, but as related to a group of other selves with whom one identifies (and who in turn mutually recognize one's identity with them) is, of course, Hegel's *Phenomenology of Mind.*

Hegel's seminal account has been supplemented by widely elaborated theoretical and experimental studies in developmental psychology, and especially in studies of child development, which focus on the emergence of the ego, and on the growth of perceptual and cognitive functions, as well as of features of personality.

Even in the case of unconscious or comatose patients, or of patients unable to communicate for other reasons (e.g., aphasias or agnosias or psychopathologies of certain sorts), we presume a self-consciousness that is, effectively (for the moment or more permanently) disabled, but nevertheless part of the individual's psychic or psycho-social constitution. Here, there arises a distinction of some practical significance in such situations. Suppose, through some injury or disease, a person suffers from a loss of conscious self-identity, in the sense that the person no longer knows who he or she is. If personal identity were simply a matter of conscious awareness of oneself as having a certain identity, e.g., knowing one's name, remembering one's situation or past experience, etc., then losing this awareness would be equivalent to losing one's personal identity. But of course, we don't ordinarily understand it in this way. We say, instead, that the personal identity remains – it is, after all, the "same" person – but that person has lost knowledge of it, for some reason. In short, personal identity would in this case seem to be an "objective" property of the individual, not dependent on the person's own awareness of it. It exists independently of that individual's consciousness. On the other hand, it doesn't seem to be simply the case that the personal identity is retained because the same bodily identity continues. Even if this bodily identity is a necessary condition for the continuity of the personal identity, it is not a sufficient condition for it. Nor does it seem to be the case that the personal identity is retained only as a social artifact, or as an ascriptive identity, resulting from the identification of the person by others, or based on records of that person. This would make personal identity a social construction, on the part of some constituting authority suitably defined – e.g., the family, the state, or the network of personal and institutional affiliations that provide the matrix of that individual's social existence. If this were in fact the way in which personal identity is determined, then it would have the odd consequence of leaving open the possibility that by a change of such ascription, the personal identity of the individual could be changed or reconstituted.

It can be plausibly argued that personal identity is not accounted for in any of these ways: it is not simply constituted reflexively, by self-awareness or self-identification; nor by ascription on the part of others or institutionally; nor by bodily identity alone. But then, what could justify the notion that personal identity remains the same, independently of all these constitutive features? Perhaps the conjunction of all these modes of identification, or of the constitution of personhood? Are they each a necessary condition, and jointly sufficient, then? Or is there some property of personal identity that human beings have (or achieve) which somehow transcends all these biological, social or psychological determinants of individual identity, though it may be rooted in them? Is personal identity an emergent property, i.e., one that derives from, but is not reducible to its component or generative elements? And if so, can it persist in the face of the elimination of its sources or causes?

These are questions that smack of some metaphysical theory of personhood or personal identity; and yet, they are posed by the requirement that our ordinary or common sense notions impose on us, and that we would be loathe to give up; namely, that personal identity is preserved in the face of loss of self-awareness of this identity, that it cannot be changed by some social or institutional choice, and that it is something more than merely having the same body continuously. We might say that we are, for the most part, epistemological realists with respect to personal identity: it exists independently of whether and how it is known (either by ourselves, reflexively, or by others). But it would be strange to conclude from this that our personal identities are only contingently tied to who we take ourselves to be, that our self-identification – i.e., some particular self-identification, as this person and not some other – is not an essential component of our "real" personal identity.

Difficulties with this arise in the case of false or mistaken identity, as well as in cases of what we may call partial loss or defect in personal identity. In a delusional state, someone may take themselves to be who they are not: a distinct other person (classically, Napoleon; more recently, Elvis), or a thing (a wall, a piece of glass), or in the case of multiple personality, several people, all of them identified with the self. The very presumption that these are delusional states presupposes that there is a "real" self or person, who is assuming these disparate and mistaken or false identities pathologically. Further, in cases of partial loss, a person may function normally in many respects, knowing what to do, how to work, where to go, and retaining all the normal characteristics of personality, with only one blind-spot: one's name, or particular identity. The notion of personal identity thus takes on a normative character, defined in terms of the well-being or healthy state of the person, in which the "right" personal identity is consciously recognized or acknowledged. Crucial to any treatment of what would then be understood to be a disorder would be a diagnosis that was concerned not only with the symptoms but also with the causes. And here, the practical question poses itself straight away: what theory of personal identity is the framework for such an inquiry? What ontology of the person is operative in this context? Willy-nilly, whether by conscious deliberation, or by tacit or unreflective adoption of one view rather than another, the search for causes to explain the identity-disorder will be guided (or constrained) by the view of what constitutes this personal identity, what it is, in effect.

V. BODY LANGUAGES AND MEDICAL PRACTICES

This brings us to the final reflection in this paper. Here, I had proposed to address the question of what difference it would make in practice to adopt one or another of the alternative ontologies of the person, or of the body, and of the relations between persons and bodies. I had also suggested that different ontological and methodological approaches may have their sources in different practices in medicine, and in alternative emphases of a reductionist or holist sort.

My conclusion is anti-climactic. I do not think it makes any difference in practice what ontology one is committed to. What makes a difference is that medical practice is a collective or cooperative enterprise – or ought to be, optimally. Whatever ontological preferences arise in the course of specialization with its attendant reductionisms, and whatever fragmentation of the patient as a whole person may arise from the division of medical labor, what saves the day is the complementarity of associated medical practices, and the resulting complementarity of alternative ontologies. If medicine were left to become, as it sometimes is, nothing more than an aggregate of different practices, which speak different "body-languages," then the object of this practice would be a patchwork of body parts, different bodies, different persons, each making their appearance on the stage cued by the particular medical occasion, the particular practitioner, the particular specialization. The unity of action in this medical "play" is that it is (in the case of the individual patient, or even in the case of a population, say, in the epidemiological case) one entity that is being treated, and this entity is a person, with a body whose parts are integral to its existence or well-being.

The various "body-languages" or "person-languages" are all about the same object, but from different perspectives, and if these remain opaque to each other, then we have a medical Babel. Alternative ontologies become exclusive rather than complementary, and for the patient, alternative diagnoses and treatments could well become mutually exclusionary. A "second opinion" would simply be another one, and so would a third. On the other hand, a complementarity of perspectives, even where differences arise which need to be resolved, enrich the cognitive matrix in which clinical judgments have to be made. And where the subject of medicine is as complex as the person, the richer the range of clinical knowledge and experience, the more apt the practice of medicine. This means, therefore, sharing of approaches, and an integrative organization of the medical practices. Medical ontology becomes a pluralistic and collaborative affair, rather than a debate among the philosophical schools.

The thoracic surgeon and the child psychiatrist have attended the same medical school, studied the same textbooks and had a large part of their medical education, both didactic and clinical, in common. True, they may have different personalities and it is conceivable that their ultimate choice of medical specialty had some sources in differences of temperament and character. But career choice is too complex a phenomenon to reduce to the simple-minded determinism of glandular or genetic causality. (I have no doubt that some of our more avid genetic determinists will soon discover the location of the thoracic surgeon gene!) Differences set in early on, in the professional focus of their respective training and experience, however. Their patients present themselves for diagnosis and treatment differently. The anatomy of the thoracic cavity and the anatomy of the mind are not easily commensurable. Yet the general consideration of the well-being of the patient, his or her ability to function normally, to be free of pain, to know what regimen to follow to avoid future medical problems, etc. – these are things all health practitioners share. If the thoracic surgeon focuses on the patient as, fundamentally, a body, as the object of a surgical procedure, he or she still needs to know the patient's medical history, and however focused upon physical information such a history may be, it also includes, essentially, information about the age, family situation, and occupation of the patient, the patient's own account of complaints or reports about symptoms – all of which is information not simply about the patient's body, or the patient's thoracic cavity, but about the person that the patient is. The surgeon may address no more than the draped body in the O.R., and in this sense, a rigorous ontological reduction of the object of the practice has taken place. But the practice, then, includes the range of practitioners who deal with this patient, prior to the surgery, during it, and afterwards. What the surgeon can ignore, or perhaps needs to ignore in the specialized focus on organs and tissues that the practice demands, becomes the responsibility of the medical team of which the surgeon is only a part. The "whole person" may not be the proximal object of the surgeon's action, whose focus is on the bodily part (in its relation to the bodily whole, of course), but the patient is the complex object of a collective or plural practice, or a set of related practices, for which the appropriate ontology is the fully elaborated person. In fact, in this optimal reconstruction, the patient is not "patient," as passive object of treatment, but co-agent, with the medical practitioners, in the process of treatment or cure. This co-agency also works at two levels: the cooperation of the body, as an organism with capacities for self-healing (which may have to be enhanced by medical or pharmaceutical intervention), and the conscious, voluntary, informed and intelligent cooperation of the person being treated, in consenting to appropriate treatment and in carrying out the regimen required for pre- or post-operative care.

A similar, but complementary story can be told about the child psychiatrist's preferred ontology. Personal identity, as a feature of psychosocial life, and of ego development or maturation, obviously suggests the choice of an ontology where self-consciousness is a principle feature of the subject. Yet child development is a matter of more than this, and pathologies of development may, in addition to their socio-cultural or family origins, be rooted in endocrine malfunction, in brain lesions, in perceptual deficiencies due to neurological causes or to dietary lacks. No less than the surgeon, the psychiatrist is a part of the whole, and only the whole is adequate to the care and treatment of the whole person. In effect, medicine, on this view, is, if not a multicultural (taking the different practices as "cultures" each with their own idioms), then certainly a multilingual enterprise.

Baruch College
City University of New York
New York, New York, U.S.A.

NOTE

[1] I deliberately do not enter here into the "bionic" fantasies that are so popular in the media, nor into the more germane issue of the integration of persons as parts of complex mechanisms, in modern warfare, or in industrial production, or in electronic information and control systems. Though conceptual issues related to those I discuss here do come up in these contexts, they are beyond the scope of this essay, and are not, properly speaking, relevant to the contexts of medical practice which are the focus here.

MARK J. CHERRY

BODIES AND MINDS IN THE PHILOSOPHY OF MEDICINE: ORGAN SALES AND THE LIVED BODY

I. INTRODUCTION

The possibility of creating a for-profit market in human organs ignites in many a feeling of deep moral repugnance. Consider, for example, physician H. Berry Jacobs, founder of International Kidney Exchange, Inc., who asked 7,500 hospitals if they would be willing to participate in his plan to broker human kidneys. He proposed to offer the poor in Third World countries and the United States the opportunity to sell a spare kidney at whatever price they would find encouraging, and then negotiate acquisition, for a profit, by Americans who could afford to pay for the organs. Jacob's plan aroused cries of moral indignation in the United States Congress, and was denounced by the National Kidney Foundation, the Transplantation Society, and the American Society of Transplant Surgeons. Professional organizations resolved to expel members who were involved in such dealings (Fox and Swazey, 1992, p. 65). As the U.S. Task Force on Organ Transplantation summarized the objection: "We are alarmed that . . . certain transplant centers are reportedly brokering kidneys from living unrelated donors. We find this practice to be unethical and to raise serious questions about the exploitation and coercion of people, especially the poor" (1986, p. 98). The argument is that the rich exploit the fact of poverty to coerce the poor into selling their organs, which given better circumstances the poor would not have done (see also Abouna *et al.*, 1990, p. 919; Dossetor and Stiller, 1990).

The moral repugnancy which is perceived to underlie commerce in human organs is held to trump the possibility of increasing life-sustaining transplants. As the Nuffield Council Report argues, certain uses of human tissue are morally unacceptable because they ". . . injure human beings by treating them as things, as less than human, as objects for use" (1995, para. 6.7). According to Russell Scott, the historical precedent for such a market is chattel slavery, in which the human body is treated as mere property (1990, p. 1003). For James Childress, such concerns

H. T. Engelhardt, Jr. (ed.), The Philosophy of Medicine, 57–78.

are grounded in basic ethical principles, such as "the dignity of the individual" and "respect for persons" (1989, p. 88). For Paul Ramsey, behind the repugnance is a real danger that such practices "will only erode still more an apprehension that man is a sacredness in the biological order . . ." (1970, p. 209). Unlike altruistically motivated donation, offering financial incentives is held to commodify persons as things and thereby to violate the dignity and respect due to persons.

Such positions are rhetorically powerful, yet they over-simplify complex issues. Whether a market in human organs violates human dignity, fails adequately to respect persons or is morally repugnant turns, in large measure, on implicit phenomenological and ontological assumptions. Among the insights which Stuart Spicker brought to the philosophy of medicine was an appreciation of the framing role which conceptions of the person and of one's relationship with one's body, play in shaping understandings of licit medical practices. For example, as Spicker points out, medical interventions, such as transplant surgery, implicitly incorporate a mechanical understanding of the body often attributed to Descartes: "it is a natural science conception, a description of a *physical* body," rather than a lived experienced body (1976, p. 117). In contrast, in everyday experience the body is lived: "we are our lived body, the body which the physician *first* encounters when we enter the clinic" (1976, p. 118). Such phenomenological observations reflect ontological understandings of the relationship between persons and their bodies, which in turn frame considerations of licit use of the body. In this essay, I critically explore such foundational issues and the implications of such considerations for the licitness of a market in human organs. As Spicker appreciated, such foundational concerns frame our understandings of the dignity and respect due to persons, which in turn inform the justification for judgments that a practice, such as organ selling, is morally repugnant.

II. NOT ALL BODY PARTS ARE CREATED EQUAL: PHENOMENOLOGICAL AND ONTOLOGICAL CONSIDERATIONS

What premises would one have to grant regarding the relation between persons and their body parts to sustain the claim that a market in human organs is morally licit? To address that question, one must begin with an account of what it is to be alive as a person regarding whom a market in organs would be of importance. Consider the ways in which persons are central for the market. Participation in the market requires that persons be beings with certain cognitive characteristics. Only persons are concerned to enter the market and to negotiate with others for fun, beneficence, or profit. It is persons who are self-conscious, self-conspicuously rational, and free to choose. Only persons claim dominion, possession, or ownership over things and self-consciously choose to cooperate with others. The most plausible characteristics for picking out those beings who are in principle capable of participating as agents in an organ market are cognitive capacities. Even to be able to have concern for ownership of or trade in human organs, much less concern for its licitness or limitations, requires that persons be the sort of beings who are able to experience self-conscious, self-conspicuous moral agency. If we grant this understanding of personhood, then beings or physical items which fail to possess these cognitive capacities cannot be affirmed as persons within the general secular context; they are things.[1] Moreover, once beings permanently loose such capacities, as Spicker points out, they are things which have the character of being former persons.[2]

Such an account carves out the phenomenologically experienced world into the moral and ontological categories of persons and things. It is persons for whom there is a possibility of self-conscious, rational stewardship, dominion, or property ownership. It is persons who are self-consciously and self-conspicuously harmed when they experience a violation of their bodily selves or other property. Only persons are able to reflect on the relative injustice of the violation. Traditional common law, for example, understood property as essentially conveying the rights of persons to exclude others from things (Hogue, 1966).[3] It expressed the rights of persons in and over things. The nature of such rights, their character, scope and form, was drawn from the nature of persons as beings who reason and will. Property has been understood as an extension of the person: this is Me, this belongs to Me, it is an extension of Me, and is protected as I am against battery, unauthorized touching (Matthews, 1995, p. 251). Indeed, under the traditional law of torts the person included any part of the body as well as anything attached to it and practically identified with it. Violation of the person included unconsented to contact with the individual's ". . . clothing, or with a cane, a paper, or any other object held in his hand . . ." (Prosser, 1971, p. 34). Interest in the integrity of the person includes his body and all the things which are in contact or connected with it. The concern is with the significance of personal integrity. As this significance increases in relevance for market decisions regarding one's personal integrity, so too will the grounds for placing the burden of proof on those who interfere with persons, and their use of their bodies, to show that they do so with moral authority.

If we grant that personhood in general secular terms requires self-conscious, self-conspicuous moral agency, then sustaining the life of a person requires only that one maintain the necessary cognitive capacities, mental and psychological, for the embodiment/existence of that particular self-conscious, self-conspicuous moral being. Consider the different ways in which self and body are both distinguishable and separable. One can distinguish but not separate those parts of the brain necessary to sustain the life of persons. One can distinguish the higher brain as physical tissue from the existence of the living person; however, one cannot separate these parts from the person. Removal of the higher brain demonstrably destroys the necessary conditions for the existence/embodiment of the person.[4] In general secular terms persons are distinguishable from all of their body parts; however, they are only separable from most. This implies that in general secular terms, a person's body, i.e., Spicker's "lived body", can be regarded as a collection of things with which the person is more or less intimately associated.

But not all things are created equal. Unlike other types of things which can be protected as part of my personal integrity (e.g., hat, cloak, or boots), body parts are experienced as me. Yet one must make distinctions even among body parts. Some parts are necessary for embodiment/existence (e.g., the higher brain); other parts are necessary for adequate human functioning (e.g., the heart); and, there are those parts which are neither (e.g., the appendix). Moreover, some parts are directly experienced (e.g., hands), while others are not (e.g., hair) (Leder, 1990, pp. 107, 111, 115). Like other types of things, though, body parts which are not necessary conditions for embodiment/existence can be replaced at will. What makes such things parts of us is being able to experience them as such.

Consider, for example, the case of a person who requires a cane to walk. When that individual purchases an improved model, or merely a new cane which suits his fancy, he replaces a thing intimately associated and experienced as part of him. He replaces a thing which is an integral part of his human functioning. Why would anyone be disturbed by the purchase of a new cane? Compare such a case with a cyborg, whose only non-human part is his left hand. The hand is experienced much like any human hand would be, with artificial nerves incorporating the extremity into his phenomenological world. If asked, he replies that he does not experience a significant difference in terms of bodily incorporation between his left hand and right hand. "It is my hand," he says, "I experience it as me." If he decided one day to shop for a new and improved hand, a stronger more adroit one, he would be seeking to replace a thing intimately associated and experienced as part of him. The hand is a body part. Yet, why would anyone be disturbed by the purchase of a new cybernetic hand? The artificial hand is experienced as part of him; however, it can none-the-less be alienated and replaced.

Similar considerations apply to xenographic organs. If we were to bio-engineer a pig to grow internal organs which have sufficiently close DNA for the organs to be transplantable into humans, the organs could be harvested and sold to persons in medical need of a transplant.[5] The new organ would replace the defective human part, eventually incorporating into and sustaining the biological life of a person. Is there a significant moral difference between the xenographic organ and the cybernetic hand? Each is engineered, artificial, and non-human in material. Both are things which we can use to replace parts of our bodies at will. Presuming adequate functioning, the organ, much like the hand, will be experienced as me once it is implanted. What ground could be advanced against a market in transplantable xenographic organs? The sale of such organs ought to be no more controversial than the sale of any other pork product.[6] What would be the moral difference, if any, between the harvesting and selling of xenographic organs and human organs?

This philosophical, fictive case is constructed in order to raise the issue of the relationship between persons and their body parts. The more the moral distance between persons and their body parts is increased, the more body parts become like other objects in the world to be possessed as well as given away or sold. A xenographic organ as fully exchangeable provides no obvious grounds against such sales. Yet is it its non-human, manufactured character that makes its sale seem plausible? The specially developed xenograft is living, but not human. Moreover, the organ is in a real sense produced or manufactured. Given these considerations, one can re-appreciate the status of those human organs which are not essential for life or consciousness. Such organs would appear more plausibly to be considered as property rather than as integral to persons. As a consequence, those arguments against a market in human organs, which are based on a reluctance to allow persons or their components to be sold would be weakened insofar as those organs are distanced from the persons who possess them. The harvesting, selling, and replacing of such parts does not harm the living experiential nature of the person.

Such considerations shift the burden of proof. Given certain understandings of body parts the permissibility of an organ market becomes more plausible. Just as one can distinguish between the sale of self and the sale of property intimately connected to oneself, which is not identical with oneself, but still experienced within the sphere of the self, one can distinguish between sale of self and sale of most body parts. Body parts which are experienced within the sphere of the self, as intimate parts of one's bodily experience, but which are not necessary for embodiment/existence in this world, can be regarded as things intimately connected to oneself, although not identical with oneself. Therefore, it is an error in reasoning to believe that one must justify a market in body parts which are needed for either embodiment/existence or adequate human functioning in order to justify a market in organs which are neither.

It is similarly an error in reasoning to object to an organ market which trades only in organs which are neither on grounds which are only concerned with embodiment/existence or adequate functioning. Equally, is an error in reasoning to conflate an argument for an organ market which trades only in organs which are required for neither embodiment/existence or adequate functioning into a market for all body parts. If the self is different from organs that are both separable and distinguishable from the self, then while the self could not sell some organs, it could sell others. For example, selling one's kidney is different than selling one's brain. While it is caught up in the life of a person, the removal and sale of a kidney will not necessarily affect either the individual's existence or embodiment as a person, or the conditions for adequate biological functioning. Unlike kidneys, the higher brain is a necessary condition for the possibility of being incarnate in this world. It is the seat of human consciousness, memory, and the cognitive capacities which sustain general secular personhood. The burden of proof is on those who oppose an organ market to show that the objections are cogent regarding the applicable body parts and the character of that part's association with persons.

Given the temporal nature of persons (i.e., they are born, live, and die), organ markets include both present and future interests. Unlike present markets in which exchange of the commodity from a living vendor takes place within an agreed time period, futures markets would allow persons to contract now for the right to harvest their organs upon death, if any remain usable. It would be an error in reasoning to presume that objections to the sale of organs which are required for embodiment/existence or adequate human functioning from living donors are equally cogent against a futures market. Futures markets do not violate constraints against killing. Moreover, willing one's organs as part of inheritance would allow the heir to profit materially through the sale while others profit physically from transplantation. Death defeats any need to sustain the conditions for either embodiment/existence or adequate human functioning, which is why much of the debate has focused on the necessary and sufficient conditions for the death of a person. At death body parts no longer sustain the life of a person; they are things. The burden of proof is on those who oppose an organ market to show that the objections are cogently focused on the characteristics of the particular market.

III. OWNING BODY PARTS AND OWNING OTHER TYPES OF THINGS

If we grant the phenomenological and ontological conclusions that persons are separable from at least most of their body parts, that parts which are separable and distinguishable from the person can be replaced without destroying either the person or the necessary conditions for embodiment/existence, and that the characteristic which marks body parts as distinct from other things is that they are experienced as part of the person, then we must further assess what it would mean to own body parts as property. Beyond a phenomenological claim regarding the first person experience of the "lived body," what would it mean to maintain that a body part is "mine" as a claim regarding dominion/possession/ownership?

Certain accounts of ownership, of the rights one has over one's property, will be more compatible with the view that one is at liberty to dispose of one's body parts as among the things one owns. An exploration of ownership must examine both the justification for different senses of ownership, as for example, in deontological right-making conditions or in certain goods or goals to be realized, as well as different senses of what it is to own an object, such as to have a steward's control over it or absolute dominion. The stronger, i.e., less contingent, ownership rights are, the more one will be forced to move along the continuum from the mere permissibility of a market in human organs to the position that such a market enjoys full *prima facie* justification. Intermediary positions will differ according to the relative strength/ contingency of ownership rights, which will in turn establish the relative burden of proof which contra-indicating considerations must meet to defeat the *prima facie* allowableness of such a market.

The range of issues can be situated along three inter-related conceptual grids. First, the closeness of the analogy between dominion/ possession/ownership of one's body parts to dominion/possession/ ownership of other types of things. The burden of proof to defeat the *prima facie* allowableness of an organ market will vary directly with the closeness of the analogy. Second, certain accounts of ownership will be more compatible with an organ market. Rights to exclusive use and to alienate one's property vary with the ground of ownership. Ownership can be freedom based, where alienation is licit as long as it is freely chosen and persons have rights to exclusive control of their property; benefit based, where alienation and exclusive use are licit only if they are integral to prudent practices or choices, conveying a benefit to oneself or others; or, special considerations based, where claims to promote special goods (e.g., equality) may defeat claims to exclusive use while alienation is illicit if it is negatively associated with the production of this special good and mandatory if alienation will positively promote it. Third, examination of necessary features of dominion/possession/ownership to which one must appeal to grant that body parts may be licitly given away, as in organ donation, while also denying the licitness of a for-profit organ market.

Consider the analogies that could be developed between dominion/ possession/ownership of one's body parts and dominion/possession/ ownership of other types of things that would make a market in organs possible. Ownership is rarely understood as forever forbidding one to rid oneself of one's property,[7] although rights to alienate, through abandonment, donation, or sale, may be only *prima facie*. It would be an inadequate conception of property to hold that while one owns the item and has exclusive use, it may never be alienated. Forbidding alienation through abandonment, donation, or sale, requires sufficiently weighty considerations to defeat the *prima facie* claims of autonomy over against possessions. Typically, such rights are fairly strong. For example, persons may generally licitly sell food and medicine, which are necessary for life, to further goods which the owner holds to be important. If owning body parts is analogous to owning other types of things, then we must look to the general secular significance of that property to persons to set the standard for justifying interference in the use of it. Insofar as the relation between persons and their body parts is held to be of greater significance than other types of person/property relationships, interfering with private use of the body must meet a higher burden of proof than interfering with private use of other owned things. As the right to be protected from interference in one's body increases, the question then arises, is it more plausible that one's right to dispose of one's body increases, or is it more plausible that it decreases? Absent contradictory considerations, it would appear that the right not to be interfered with should increase one's property rights in one's body.

How one proceeds further to determine what would support a market in organs depends on numerous and complex considerations. If one may sell oneself, then one may sell one's organs. For example, if one can join the military and set one's life at risk, does it follow that one can set one's body slightly at risk through the sale of body parts? Even if we discount voluntary slavery and donation/sales that would involve a loss of life, if the self is different from the organs that are separable and distinguishable from the self, though the self may not sell some organs, it may sell others. Just as one only requires sufficient food and medicine to sustain life, rather than any particular piece of food or dose of medicine, of those body parts which are both distinguishable and separable from the self, one only requires a sufficient set of body parts, rather than any particular parts or replacements, to sustain the biological life which in turn sustains personal existence/embodiment. Even replacing one's heart or liver with an adequately functioning equivalent will preserve the existence/embodiment of the person. Indeed, body parts which are distinguishable and separable may be the best example of body parts which can be most easily understood on an analogy with external possessions. Such organs are things which can in principle be exchanged. This suggests that of all the things persons own, they have the strongest rights for exclusive use, and thus for alienation, in their body parts, which in turn sets the standard of proof for interference with personal use higher than other types of property. Here, more must be said regarding what circumstances might make certain possessions inalienable beyond transfer or sale. Here, too, one will need to explore what conditions would forbid the sale of an organ when it could be given as a gift.

The significance, or strength, of rights to exclusive use and to alienate one's property also vary with the ground of ownership. If ownership is grounded on autonomy based rights, for example, then alienation, through abandonment, donation, or sale, would be *prima facie* licit as long as freely chosen. Along with association and occupation, economic freedoms are fundamental elements of total freedom (Friedman, 1982, p. 77). Owners have exclusive use of their property and among the pursuits to which it may be put is the creation of profit. Things which one owns, including body parts, would then be open to be sold as long as both vendor and purchaser freely agree to the transaction. Other values or special considerations, such as benefits to others or feelings of moral repugnance, may not easily meet the burden of proof licitly to forbid sales. Considerations of benefits to others may not defeat autonomy based rights.

Coercion by need or other external circumstances may similarly fail to meet the burden of proof to nullify the agreed to transaction. As external conditions are admitted to defeat the claim freely to have sold or purchased property they will also destroy the possibility of a market in any product, much less in body parts. Every transaction will be subject to mutual reciprocal claims of external coercion. For example, a poor person who sells a kidney to a rich patient may argue that he was coerced into selling by poverty. The rich patient, though, may equally argue that he was coerced into purchasing by kidney failure (Bayles, 1974; Benditt, 1978; Gorr, 1986). Defeating the licitness of the transaction requires demonstrating that one of the parties coerced the other or that the circumstances rendered consent impossible. In this case, the parties are not directly in the relationship of coercer and coerced. Nor are there grounds to hold that the circumstances of the external constraints of nature would render either party incompetent to choose. Other values may be offered as reasons peaceably to persuade persons not to enter into certain types of transactions, to motivate different choices, but freedom should in general trump. That is, much more would need to be made out in order to show that valid agreement would not be possible. Ownership based on freedom is compatible with the view that one is at liberty to dispose of one's body parts as among the things one owns: property rights, including exclusive use and alienation, are actual rather than *prima facie*. Such an account would be sufficient to support the position that a market in human organs could not be rightfully forbidden.

If ownership is benefit based then rights to exclusive use or to alienate property is licit only if prudent. Rules regarding property rights are only justified insofar as they convey benefit to oneself or others. One manner in which to understand prudence is as maximizing overall benefits while minimizing harms. Consider utilitarian property theory, for example: a system of ownership is only justified if it maximizes the good (Singer, 1981). Or, as with John Rawls, when it maximizes the greatest good for the greatest number (1971; 1993). As David Hume pointed out, scarcity of resources may cause societal friction and violence as individuals compete for resources. A carefully designed property system ought to promote social stability and peace, minimizing violent competition for the same resources, thereby maximizing personal and social good.[8] Everyone will be better off because each will be left in the peaceful enjoyment of those items acquired through luck or hard work. Moreover, private ownership may be a necessary condition for efficient resource usage. Without rights to exclusive use and alienation there is no reasonably assured reward for incurring the costs of developing materials and investing time and labor. For example:

> A farmer plants corn, fertilizes it, and erects scarecrows, but when the corn is ripe his neighbor reaps and sells it. The farmer has no legal remedy against his neighbor's conduct since he owns neither the land he sewed nor the crop. After a few such incidents the cultivation of land will be abandoned and the society will shift to methods of subsistence (such as hunting) that involve less preparatory investment (Posner, p. 27).

Parceling out rights to exclusive use and to alienate for a profit can cultivate incentives for maximizing the benefits gleaned from available resources. Here, with respect to benefits-orientated understandings of ownership and their implications for markets in organs, one should note two quite different foci in terms of which to calculate benefits: (1) how best to use the particular organs (e.g., will a market lead to a better use of organs), and (2) how best to maximize benefits for persons. The second question will depend on the first. It will always be important, given the benefits of organ transplantation, to examine the first question in order to be able appropriately to address the second.

Exclusive use of body parts, on such an account, is contingent on such rights conveying the appropriate benefit, such as efficient use of resources or maximizing the good. This means that the distribution and use of internal organs is fixed neither by birth, nor by gift or sale, but in terms of an *a posteriori* calculation of benefits. Such an account of property is in certain ways compatible with an organ market. Insofar as such a market maximizes benefits over harms it is allowable, indeed commendable. Confirming rights to exclusive use of one's body parts may efficiently and effectively produce the most beneficial uses of the available organs. Those who would not otherwise donate might be encouraged to sell their internal redundant organs, thereby saving lives and increasing personal wealth. Forbidding such transactions, though, would only require defeating the claim that it conveys a sufficient benefit. Concerns may be raised, for example, about the sale of poor quality or diseased organs. Such concerns are justified only if no adequate system of personal liability or independent methods to monitor quality can be developed. Moreover, assuming adequate long term functioning, a dying patient may prefer transplantation of a diseased organ to death in the immediate future. Nevertheless, *prima facie* rights to exclusive use and alienation of body parts are straightforwardly defeated by demonstrating that other uses, distributions, or methods of alienation convey greater benefit.

Regardless of whether one is calculating across preference satisfaction, wealth, or welfare maximization, the strength of actual ownership rights, and thus the burden of proof to defeat such rights, must be settled by empirical calculation. Depending on the results, alienation of internal organs for redistribution may not simply be permissible but required. If a policy of redistributing working kidneys from those who have two to those with kidney failure would produce sufficient benefits then this would constitute sufficient grounds to redistribute. While forced redistribution is incompatible with the buying and selling of organs, in itself it commodifies organs as resources to be utilized to maximize benefits. Here, it is important to review the different senses of commodification at stake: commodities are not just items which are bought and sold for money, they may also be resources traded among persons to produce conditions or social circumstances which are valued, which can be understood in terms of the production of maximum overall benefits, or of the greatest benefit for the greatest number of persons. On such a foundation, persons will only have ownership rights in body parts insofar as it is prudent for society to support such rights. The outcome of such empirical calculations may or may not support the view that one is at liberty to dispose of one's body parts as one may the things one owns.

Accounts of ownership which focus on special considerations argue that certain goods have a trumping character when determining licitness of ownership rights to exclusive use and alienation. Consider, for example, egalitarians who argue that to be just society must be structured, including personal holdings and the private use of property, so that it promotes fair equality of opportunity. To justify even private transactions, one must demonstrate that they do not violate the necessary conditions for fair equality of opportunity (e.g., Daniels, 1976; 1996). For example, if it is claimed that fair equality of opportunity requires substantively equal legal and political liberty, which in turn is dependent on a sufficient level of economic equality, then insofar as actual distribution of property fails to be adequately equal, there must be further redistribution. Depending on how conditions for fair equality of opportunity are specified, citizens may be entitled to substantively equal material or economic conditions, educational opportunities, or health care. Providing for such positive entitlements requires taxation and redistribution of the time, talent, goods and resources of other people, defeating claims to have exclusive use of oneself or one's property. Such ownership rights will tend to be much weaker and more contingent than those based on freedom, depending on whether considerations of free choice establish inviolable side-constraints as well as how highly freedom is valued.

Arguments which appeal to the trumping character of equality, or some other special good, attempt to defeat claims of exclusive use of property. The idea of property is recast in terms of a social benefit to be redistributed. Such a theory moves the burden of proof away from governmental and other authorities, who would continually interfere with owners' freedom, placing it on those persons who would freely utilize private property with consenting others. Rights to alienation, for example, are defeated if exercise of such a right is detrimental to the production or maintenance of this special good. Consider, for example, Nozick's case of "Wilt Chamberlain":

> . . . suppose that Wilt Chamberlain is greatly in demand by basketball teams, being a great gate attraction . . . He signs the following sort of contract with a team: In each home game, twenty-five cents from the price of each ticket of admission goes to him . . . The season starts, and people cheerfully attend his team's games; they buy their tickets, each time dropping a separate twenty-five cents of their admission price into a special box with Chamberlain's name on it. They are excited about seeing him play; it is worth the total admission price to them. Let us suppose that in one season one million persons attend his home games, and Wilt Chamberlain winds up

> with $250,000, a much larger sum than the average income. . . . Is he entitled to this income (1974, pp. 161–162)?

If the special good which underlies ownership rights is equality, then, in this case, the free alienation of the $0.25 per person is illicit since taken as a group such activity is negatively associated with the production of equality.[9] Importantly, however, alienation may be mandatory if abandonment, donation, or sale will positively promote the good.

Egalitarian political theory, for example, is often concerned with correcting inequalities created by natural and social circumstances (see Rawls, 1971, p. 74). Such considerations, though, implicitly commit these theories to a market in body parts. Among the most prevalent and pervasive inequalities, which also lead to social and economic inequalities, are those created by the differences in the health status of one's body parts. Indeed, such considerations refocus concern away from whether to commodify internal redundant organs to an account of the fair distribution of body parts (e.g., redistributive justice of physical resources) and the fair distribution of economic and social burdens to pay for acquisition, allocation, and transplantation (e.g., redistributive justice of social and economic resources). Concerns to promote fair equality of opportunity may entitle persons with organ failure to governmentally supported transplant surgery, including follow-up care and provision of a new organ. These considerations might lead to a blending of egalitarian and liberty-oriented concerns with respect to the development of a state-controlled market in body parts. For example, one might imagine a circumstance under which the price of body parts in the market would be fixed by the government, and the poor provided with body part vouchers. Such a circumstance would not allow the rich to outbid the poor in the market. There would be the additional virtue of the poor having vouchers at their disposal to act as an incentive to bring individuals to sell organs to those in need. Other regulatory impositions could be fashioned toward the goal of serving other societal interests.

IV. THE LIVED BODY, HUMAN DIGNITY, AND MORAL REPUGNANCE

Regardless of the grounds of ownership rights, some will insist that the analogy between dominion/possession/ownership of body parts and dominion/possession/ownership of other things breaks down at the point of for-profit transactions. The argument is that it is precisely because of the general significance and intimate connection with persons that organs ought not to be sold. The body is part of the basic dignity of the human person, therefore, trade in the body and its parts is morally repugnant and ought to be limited.

As a report from the Hastings Center states:

> The view that the body is intimately tied to our conceptions of personal identity, dignity, and self-worth is reflected in the unique status accorded to the body within our legal tradition as something which cannot and should not be bought or sold. Religious and secular attitudes make it plain just how widespread is the ethical stance maintaining that the body ought to have special moral standing. The powerful desire to accord respect to the dignity, sanctity, and identity of the body, as well as the moral attitudes concerning the desirability of policies and practices which encourage altruism and sharing among the members of society produced an emphatic rejection of the attempt to commercialize organ recovery and make a commodity of the body (1985).

Similarly, Leon Kass argues that there is a general repugnance at the notion of owning living nature *per se* (1985; see also Kass, 1997). As Paul Ramsey holds, the human body has a sacredness in the biological order. "In terms of our vision of man and his relation to community, there may be little to choose between the blood and soil, organic view of the Nazi and the technological, '(spare parts)' mechanistic analogies of the present day" (1970, p. 168). In each case, it is an appeal to a moral repugnance apparently felt by some in the case of organ sales.

Here, however, one must adjudicate among moral intuitions. One must distinguish between justified and unjustified moral repugnance. For example, many have deep intuitions regarding the moral repugnance of abortion, contraception, and homosexuality, yet these are practices which society permits. Is such repugnance justified? If so, should such practices be prohibited? *Pace* Kass, who asserts the moral wisdom of repugnance, in the case of organ sales, one must determine whether generalized feelings of moral repugnance are justified, prior to presupposing that such intuitions ought to carry any weight in meeting the burden of proof, or lowering the standard of proof, to proscribe organ sales.

Moreover, insofar as such theorists want to allow for morally licit organ donation from living or deceased donors, they must appeal to some special value regarding human organs to support this normative claim. Such values must appeal to "concerns about the inappropriateness of commerce in vital human organs, concerns about justice, a feel for the awesome nature of such gifts, and some uneasiness about shutting families out of the process" (Murray, 1996, p. 111), or other similar interests. Yet, not all share such unease. Feelings about the inappropriate nature of commerce in human organs can be countered by equally strong but contrary feelings. Deep intuitions regarding the awesome nature of the "gift" beg the question regarding the meaning, significance, and importance of internal organs (Kahan, 1996; Barkan, 1996). The nature and outcome of the "gift" are no less "awesome" if money changes hands. Those who are concerned about the "dirtiness" of such transactions need not participate. Vigorous searching for donors need not cease merely because a for-profit market is licit. General concerns about shutting out the family must be dealt with regardless of the mechanism for the acquisition of organs.

V. CONCLUSION

The foregoing reflections disclose, first, the possibility of driving a conceptual wedge between persons and their body parts. Insofar as persons are self-conscious, rational, embodied beings with a view of the good, integrity depends on some but not all of their organs. Those organs that are necessary conditions for their being embodied in the world will possess a unique status. Selling them will be tantamount to selling or killing persons. Insofar as arguments against the sale of organs depend on "respect for persons", such arguments will govern; however, they will not hold over those organs which are not essential for the continued life of persons. The more the organs involved can be replaced without a loss in the range of embodied function usual to humans, the less plausible will there be a claim that they should be considered as different from other property that can be given or sold. The more body parts are like other replaceable objects (glasses, canes, or cybernetic hands), the more plausible it becomes that they should be open to being bought and sold in the market. A necessary condition for the legitimacy of a market in organs will be that the organs bought and sold are not integral to persons as such, but can be conceived of as outside of the essential core of their embodiment.

Second, persons are more intimately associated with their bodies than they are with other types of things. This phenomenological and ontological consideration provides significant reason for accepting the further conclusion that interfering with private use of the body, and of one's body parts, must meet a higher burden of proof than interfering with private use of other owned things. General secular rights not to be interfered with increase the significance of the rights persons have to exclusive use and control in their bodies. Insofar as persons may licitly sell other types of things, it is plausible to assume that it is licit to alienate for a profit at least those body parts which are not necessary for adequate human functioning. Whether ownership is grounded in personal autonomy or social benefit, a market in human organs may better respect persons and protect their dignity as rational, self-conscious, moral beings, than alternative strategies for procuring and allocating organs for transplantation.

Finally, all commodify, even donation. That is, on either ground one has specified a market in human organs, albeit a heavily regulated market, with carefully stipulated conditions for bearing the costs and benefits of procurement, distribution, and transplantation. One is not arguing about whether kidneys should be commodified but over who should receive the resources and who should bear the costs of appropriation and transfer. Insofar as individuals are prohibited from selling their organs, it is a constrained market where donors are required to part with their property without material compensation, while others including physicians, hospitals, and procurement agencies benefit financially, and the recipient of the transplant benefits physically. That organs can only be transferred at a price of zero does not thereby reduce the value of such organs to zero. Rather, it transfers the value of the organ from the donor to other parties (OTA, 1987, p. 124).[10] Even if mechanisms for distribution of internal organs are strictly regulated, non-capitalistic, non-profit driven, governmentally financed markets, organs are none-the-less socially traded commodities, utilized for production or maintenance of special social goods. Therefore, there are good grounds for concluding that the particular moral repugnance reserved for markets in which organ donors, or their families, profit is unjustified.

Together such considerations provide significant grounds for concluding that a market in human organs respects persons and preserves human dignity more adequately than other strategies for human organ procurement and allocation for transplantation, and, moreover, that such a practice is not plausibly understood as justifiably morally repugnant.

Department of Philosophy
Saint Edward's University
Austin, Texas

NOTES

[1] Membership in the species *homo sapiens* is generally appreciated as insufficient for a being to be a person. Cognitive capacities are the most plausible candidates for characteristics which distinguish persons from things. For example, Aquinas stated that "Person signifies what is most perfect in all nature – that is, a subsistent individual of a rational nature" (*ST*, I, Q29, a3). John Locke identified persons as beings with the capacity of self consciousness coupled with a rather rudimentary intelligence. A person is a "thinking intelligent being, that has reason and reflection, and can consider itself as itself, the same thinking thing in different times and places, which does only by that consciousness which is inseparable from thinking, and, as it seems to me, essential to it" (*Essays*, book II, chapter xxvii).

Kant marked the distinction as: "A *person* is a subject whose action can be *imputed* to him. . . . a *thing* is that to which nothing can be imputed. Any object of free choice which itself lacks freedom is therefore called a thing (*res corporalis*) (1991 [1797] §223). Similarly, Hegel argued that "person" is a category which captures that group of beings capable of conceptualizing and embodying a self understanding in the world (1967 [1821] §42). In contrast, "Since a thing lacks subjectivity, it is external not merely to the subject but to itself . . . i.e., that whose determinate character lies in its pure externality" (§42A). Unlike things, Hegel argued, persons conceptualize themselves, shaping their own actions and characters to a significant extent.

Contemporary philosophers Michael Tooley and Ronald Dworkin cast the distinction respectively as "entities that have a right to life" (1984, pp. 120–134) and as "human beings who are capable of forming and acting on intelligent conceptions of how their lives should be lived" (1977, p. 272). See also Harris 1992, chapter one; 1995, pp. 6–21; Nozick, 1981, p. 78; Engelhardt, 1993, p. 420; Dennett, 1988; Brody, 1974; Callahan, 1988; Becker, 1975.

Religious understandings of personhood also often refer to cognitive capacities: "From a Christian perspective the most basic requirement for 'personhood' is the unity of body and soul. However we may define 'soul' it is clearly related to brain function. That is, once brain death occurs, the organism is dead . . . The gift of human life depends, rather, upon the cerebrum (left and right hemispheres) and its covering of nerve cells known as the cortex. The cortex enables us to make decisions, organize our lives, remember experiences, communicate with others, and perform various creative activities, such as art and music. It is here as well that our capacity for reflective activities resides, e.g., the ability to philosophize, to speculate, to wonder, and to pray" (Breck, 1995, p. 332; see also, Ramsey, 1970; Crosby, 1996).

For an account which attempts to reject cognitive capacities as the mark of personhood see Meilaender, 1995.

[2] As Spicker points out, bodies of the recently deceased cease to have the status of a person: "The neomort may in fact be appealing to those in medicine who are intent on prolonging life, not simply the 'life' of the neomort, but the life of the recipient person as well. But what precisely is a neomort? More importantly, perhaps, what do we mean by that body?" (1976, pp. 116–117).

[3] For example, the protection of possessions was at the center of Henry II's legislation between 1164 and 1179 (Hogue, 1966, pp. 161–163). The common law of the 12th and 13th centuries was in large measure the law of property rights (p. 112). In the 19th century,

Blackstone, reflecting on the common law of England, argued: "There is nothing which so generally strikes the imagination and engages the affections of mankind, as the right of property; or that sole and despotic dominion which one man claims and exercises over the external things of the world, in total exclusion of the right of any other individual in the universe" (Blackstone, 1803, book 2, p. 1; See also Buckle, 1991).

[4] Or as Puccetti put it, where the brain goes, there too goes the person (1969, p. 65).

[5] Currently, xenografts have not been very successful. For example, in 1984 a surgeon transplanted the heart of a baboon into an infant whose heart lacked a left ventricle. The girl died several days later (Altman, 1984; see also Jonsen, 1997, p. 253; Sells, 1994).

[6] For opposing views regarding animal rights and xenotransplantation see Francione, 1990; Reemtsma, 1990.

[7] Samuel Pufendorf, for example, characterizes ownership as "a right, by which what one may call the substance of a thing belongs to someone in such a way that it does not belong in its entirety to anyone else in the same manner" (1991 [1672], p. 85). It follows, he adds, that "we may dispose as we will of things which belong to us as property and bar all others from using them, except insofar as they may acquire a particular right from us by agreement" (1991 [1672], p. 85). See also Becker, 1980; Honoré, 1961.

[8] ". . . [W]hen they have observ'd that the principal disturbance in society arises from those goods, which we call external, and from their looseness and easy transition from one person to another; they must seek a remedy . . . this can be done after no other manner, than by a convention enter'd into by all the members of the society to bestow stability of the possession of those external goods, and leave everyone in the peaceful enjoyment of what he may acquire by his fortune or industry" (cited in Buckle, 1991, p. 489).

[9] See Cohen (1977) for an argument concerning the ways in which enforcing such patterning of economic distribution preserves liberty.

[10] The Office of Technology Assessment considered the following case in their report on the ownership of human tissues and cells: "Suppose, for example, that a market for transplantable kidneys existed. There are three parties to the transaction – the donor, the surgical/hospital team, and the recipient – and they are able to accomplish the transaction at prices agreeable to all. The amount required by the family of the kidney donor to proffer the kidney is $50,000, the amount required by the surgical/hospital team to bring forth its services is $100,000, and the kidney recipient is willing to pay $150,000. Suppose further that a law is passed requiring all transactions in kidneys to be gifts, thereby prohibiting the kidney donor's family from selling the kidney and reaping its economic value of $50,000. Who will now realize this value? The intent of the legislation was that the value of the kidney be transferred as a gift from the kidney donor to the recipient, with the transplant ultimately costing the recipient only $100,000. Yet, because nothing is done to ensure this outcome, a different outcome is possible. Depending on the conditions of the transplantable kidney market, it may be possible for the surgical/hospital team to realize the value entirely by charging the recipient $150,000" (1987, p. 124).

BIBLIOGRAPHY

_____: 1985, *Ethical, Legal, and Policy Issues Pertaining to Solid Organ Procurement: A Report to the Project on Organ Transplantation*, The Hastings Center, New York.

Abouna, G. M. *et al.*: 1990, 'Commercialization in Human Organs: A Middle Eastern Perspective', *Transplantation Proceedings* 22, 918–921.

Altman, L. K.: 1984, 'Learning from Baby Fae', *The New York Times* (Nov. 18), 1, 30.

Aquinas, T.: 1947, *Summa Theologia*, Benzinger Brothers, New York.

Barkan, L.: 1996, 'Cosmas and Damian: Of Medicine, Miracles, and the Economics of the Body', in S. Youngner *et al.* (eds.), *Organ Transplantation: Meanings and Realities*, The University of Wisconsin Press, Madison.

Bayles, M.: 1974, 'Coercive Offers and Public Benefits', *The Personalist* 55.

Becker, L.: 1975, 'Human Being: The Boundaries of the Concept', *Philosophy and Public Affairs* 4, 4 (Summer).

Becker, L.: 1980, 'The Moral Basis of Property Rights', in J. R. Pennock and J. Chapman (eds.), *Property*, New York University Press, New York, pp. 187–220.

Benditt, T.: 1978, 'Threats and Offers', *The Personalist* 58.

Blackstone, W.: 1803, *Commentaries on the Laws of England*, G. Tucker (ed.), Augustus, M. Kelley Publishers, New York; Rothman Reprints, Inc., South Hackensack, NJ.

Breck, J.: 1995, 'Euthanasia and the Quality of Life Debate', *Christian Bioethics* 1, 322–337.

Brody, B.: 1974, 'On the Humanity of the Fetus', in R. L. Perkins (ed.), *Abortion: Pro and Con*, Schenkman Publishing Co., New York.

Buckle, S.: 1991, *Natural Law and the Theory of Property: Grotius to Hume*, Clarendon Press, Oxford.

Childress, J.: 1989, 'Ethical Criteria for Procuring and Distributing Organs for Transplantation', in J. Blumstein and F. Sloan (eds.), *Organ Transplantation Policy: Issues and Prospects*, Duke University Press, Oxford.

Cohen, G. A.: 1977, 'Robert Nozick and Wilt Chamberlain: How Patterns Preserve Liberty', *Erkenntnis* 11, 5–23.

Crosby, J.: 1996, *The Selfhood of the Human Person*, Catholic University Press, Washington, DC.

Dennett, D.: 1988, 'Conditions of Personhood', in M. Goodman (ed.), *What is a Person?* Humana Press, Clifton, NJ, pp. 145–168.

Dossetor, J. B. and Stiller, C. R.: 1990, 'Ethics, Justice, and Commerce in Transplantation', *Transplantation Proceedings* 22, 892–895.

Dworkin, R.: 1977, *Taking Rights Seriously*, Harvard University Press, Cambridge.

Engelhardt, H. T., Jr.: 1993, 'Personhood, Moral Strangers, and the Evil of Abortion: The Painful Experience of Post-modernity', *The Journal of Medicine and Philosophy* 18, 419–421.

Fox, R. and Swazey, J.: 1992, *Spare Parts: Organ Replacement in American Society*, Oxford University Press, New York.

Fracione, G.: 1990, 'Xenografts and Animal Rights', *Transplantation Proceedings* 22, 1044–1045.

Friedman, M.: 1982 [1962], 'Freedom under Capitalism', in T. R. Machan (ed.), *The Libertarian Reader, Rowman and Littlefield*, Totowa, NJ, pp. 76–85.

Gorr, S.: 1986, 'Toward a Theory of Coercion', *Canadian Journal of Philosophy* 16, 383–406.

Harris, J.: 1992 [1985], *The Value of Life*, Routledge, London.

Harris, J.: 1995, 'Euthanasia and the Value of Life', in J. Keown (ed.), *Euthanasia Examined*, Cambridge University Press, Cambridge.

Hegel, G. W. F.: 1967 [1821], *Philosophy of Right*, T. M. Knox (trans.), Oxford University Press, New York.

Hogue, A.: 1996, *Origins of the Common Law*, Liberty Press, Indianapolis.

Honoré, A. M.: 1961, 'Ownership', in A. Guest (ed.), *Oxford Essays in Jurisprudence*, Oxford University Press, Oxford, pp. 370–433.

Jonsen, A.: 1997, 'Ethical Issues in Organ Transplantation', in R. Veatch (ed.), *Medical Ethics*, second edition, Jones and Barlett Publishers, Sudbury, MA.

Kahan, B.: 1996, 'Organ Donation and Transplantation – A Surgeon's View', in S. Youngner *et al.* (eds.), *Organ Transplantation: Meanings and Realities*, The University of Wisconsin Press, Madison.

Kant, I.: 1991 [1797], *The Metaphysics of Morals*, M. Gregor (trans.), Cambridge University Press, Cambridge.

Kass, L.: 1985, *Toward a More Natural Science*, Free Press, New York.

Kass, L.: 1997, 'The Wisdom of Repugnance', *The New Republic* (June 2), 17–25.

Leder, D.: 1990, *The Absent Body*, Chicago University Press, Chicago.

Locke, J.: 1977 [1690], 'An Essay Concerning Human Understanding', in S. Cahn (ed.), *Classics of Western Philosophy*, second edition, Hackett Publishing Co., Indianapolis.

Matthews, P.: 1995, 'The Man of Property', *Medical Law Review* 3, 251–274.

Meilaender, G.: 1995, *Body, Soul, and Bioethics*, Notre Dame University Press, Notre Dame.

Murray, T.: 1996, 'Organ Vendors, Families, and the Gift of Life', in S. Youngner *et al.* (eds.), *Organ Transplantation: Meanings and Realities*, The University of Wisconsin Press, Madison.

Nozick, R.: 1974, *Anarchy, State, and Utopia*, Basic Books, New York.

Nozick, R.: 1981, *Philosophical Explanations*, The Belknap Press of Harvard University Press, Cambridge.

Nuffield Council on Bioethics: 1995, *Human Tissue: Ethical and Legal Issues*, London.

Office of Technology Assessment: 1987, *New Developments in Biotechnology: Ownership of Human Tissues and Cells*, U.S. Government Printing Office, Washington, DC.

Posner, R.: 1986, *The Economic Analysis of Law*, third edition, Little, Brown and Co., Boston.

Prosser, W.: 1971, *Law of Torts*, West Publishing Co., St. Paul.

Puccetti, R.: 1969, 'Brain Transplantation and Personal Identity', *Analysis* 29, 65.

Pufendorf, S.: 1991 [1672], *On the Duty of Man and Citizen According to Natural Law*, J. Tully (ed.), M. Silverthorne (trans.), Cambridge University Press, Cambridge.

Ramsey, P.: 1970, *The Patient as Person*, Yale University Press, New Haven.

Rawls, J.: 1971, *A Theory of Justice*, Harvard University Press, Cambridge.

Rawls, J.: 1993, *Political Liberalism*, Columbia University Press, New York.

Reemtsma, K.: 1990, 'Ethical Aspects of Xenotransplantation', *Transplantation Proceedings* 22, 1042–1043.

Scott, R.: 1990, 'The Human Body: Belonging and Control', *Transplantation Proceedings* 22, 1002–1004.
Sells, R. A.: 1994, 'Transplants', in R. Gillon, P*rinciples of Health Care Ethics*, John Wiley & Sons, New York, pp. 1003–1025.
Spicker, S. F.: 1976, 'Terra firma and Infirma Species: From Medical Philosophical Anthropology to Philosophy of Medicine', *The Journal of Medicine and Philosophy* 1, 104–135.
Singer, P.: 1981, 'The Right to be Rich or Poor', in J. Paul (ed.), *Reading Nozick: Essays on Anarchy, State, and Utopia*, Rowman & Littlefield, Totowa, NJ.
Tooley, M.: 1984, 'In Defense of Abortion and Infanticide', in J. Feinberg (ed.), *The Problem of Abortion*, Wadsworth Publishing Co., Belmont, pp. 120–134.
United States Task Force on Organ Transplantation: 1986, *Organ Transplantation: Issues and Recommendations*, U.S. Department of Health and Human Services, U.S. Government Printing Office, Washington, DC.

SECTION TWO

EUTHANASIA, SECULAR PRIESTS, AND THE CENTRALITY OF CHOICE

THOMAS HALPER

ACCOMMODATING DEATH: EUTHANASIA IN THE NETHERLANDS

I. INTRODUCTION

Years ago, when Stuart Spicker and I were seeking to detour his younger son around a four stage bereavement over the unexpected demise of his obese guinea pig, we chanced upon a children's magazine filled with "What is wrong with this picture?" games. The cartoons were not very subtle – a fish sitting on a sofa watching television, a boot hanging from a fruit tree – their appeal being more to a child's fondness for dopeyness than an adult's for paradox. But the magazine served its purpose by changing the subject, and we were all grateful for it.

A quarter century later, I cannot avoid raising the same question: What is wrong with this picture? *A festschrift* for Stuart Spicker? Of course, he is worthy. It takes two hands merely to lift his *curriculum vitae*, and everyone knows that the contemporary flourishing of philosophy in its medical applications would have been quite impossible without his innumerable contributions. Yet this *festschrift* appears very premature. *Festschrifts* normally signal the end of careers, and yet Spicker still has so much to teach us and so many years in which to do his teaching. And euthanasia as a *festschrift* topic? With his legendary, voracious appetite for life, Spicker seems an odd person to honor with a discussion of how a society deals with death.

The oddness, however, is only superficial, for Spicker has always known that death and dying are parts of life – indeed, among the most mysterious and unfathomable parts – and that is why he has often selected them as subjects for his characteristically clear-eyed analysis. Consider this passage, for example, which not only empties a rhetorical Augean stable but leaves its floors clean enough to eat on:

> a "death with dignity" is not equivalent to a death unbetubed and without artificial devices but a death that proceeds according to the *will* of the patient. The undignified death is the death which ensued on the basis of the values of persons who are *not* authorized to speak for the patient. When we override the freedom of any patient we risk the foundation on which dignity rests (Spicker, 1979, p. 44).

A more succinct statement of the issue is hard to conceive.

H. T. Engelhardt, Jr. (ed.), The Philosophy of Medicine, 81–116.

Spicker's concerns, however, have never been wholly abstract. He is simply too much a man of the world to tolerate separation from it. His analysis of the execution of euthanasia, for instance, is deeply rooted in the experience of real patients and real physicians, and it is this quality that gives it its special power (Spicker, 1978). But for Spicker it is perhaps the Dutch context that has long exerted its strongest pull, for he has spent considerable time there, developing strong personal and professional ties and a profound affection and respect for the Dutch people and their fascinating and under-appreciated culture. How the Netherlands confronts a practice posing a life-or-death issue for thousands of its citizens every year, therefore, is a subject that would instantly engage Spicker's attention.

The preeminent contemporary analyst of Dutch politics undoubtedly is Lijphart (Lijphart, 1968, 1971, 1975), who found the Netherlands a society of reinforcing and not cross-cutting cleavages. Thus denied a widely assumed prerequisite for stable democracy (Coser, 1956; Dahrendorf, 1959; cf., Lipset, 1960), the Netherlands persisted in its stable democracy, anyway. How to explain this apparent paradox? Lijphart's answer built on the Dutch word *verzuiling*, a term derived from *zuil* (pillar) and meaning, literally, pillarization. The image suggested is a house, whose roof is supported by pillars that are independent of each other and yet part of and supportive of the larger structure. So it was with the Netherlands, a profoundly divided, "segmented society [with] tightly knit systems of organizations" (Irwin, 1988, pp. 82, 83). Dutch society was seen as divided into more or less autonomous rival blocs, each able pretty much to govern its own affairs. Religious and class divisions ran deep, and even relatively minor blocs sometimes had their own economic organizations, media outlets, and even political parties. Though *verzuiling* in its modern form may be traced to an acrimonious battle in 1857 over government recognition and subsidization of religious schools, the roots of the practice go back to the sixteenth century.

Each pillar was capped by a collection of leaders, authoritarian toward their followers yet moderate in their political views. Foreswearing much of what other democratic societies accepted as ordinary intergroup competition, these leaders preferred to come together to resolve conflicts among themselves or to let matters drift, if that appeared the wiser course. Thus, though their constituents may have been ideologically stubborn, this obstinacy was negated by the leaders' commitment to a highly pragmatic, elite dominated game that they viewed as indispensable to the maintenance of their democratic system; "The centrifugal tendencies inherent in a plural society are counteracted by the cooperative attitudes and behavior of the leaders of the different segments of the population" (Lijphart, 1977, p. 1).

Leader cooperation in such a system was facilitated by formally incorporating its chief divisions into its major political institutions. The cabinet (and bodies of lesser importance, like advisory commissions) contained representatives from various blocs, public funds and the appointment of public officials followed at least a rough proportionality, and the blocs retained substantial control over their own affairs (including such matters as education).

This system, sometimes called "consociational" after Johannes Althusius's concept of *consociatio* in his *Politica Methodice Digesta* (Althusius, 1932), was reinforced by several traditions: a multiparty system lacking a majority party, so that no single bloc could dominate all the others; easily identifiable blocs, each with a respected leader to advocate its interests; norms under which elites were expected to dominate the public but compromise among themselves; a public that is generally passive, deferential, and accepting of the modest role assigned to it; family and interpersonal relationships that perpetuate distinctions among blocs; and a very widespread loyalty among leaders and the public to the nation as a whole and its legally constituted democratic system. Lijphart dubbed the resulting politics "the politics of accommodation" (an historian preferred "pacificatory democracy" [Kossman, 1978, p. 569]) and later scholars have found similar phenomena in several other small Western European settings (Nordlinger, 1972; Powell, 1970; Steiner, 1974).

While relying more on historical narrative than hard edged constructs, Lijphart and his disciples nonetheless were clear in emphasizing the role performed by four principal *zuilen* (Catholics, Calvinists, liberals, and socialists). Fragmentation followed these highly organized ideological-religious blocs, whose cooperation and competition governed the Netherlands' political life. The corollary was that as the bases of these blocs began seriously to erode in the 1960s, the era of *verzuiling* has "been declining since then" (Lijphart, 1977, p. 2). Lijphart himself pointed to less salient religious cleavages, greater interactions among members of different *zuilen* (Lijphart, 1971), more weakly defined social differences, declining public docility, and the growing predilection of leaders to reject the imperative to work together (Lijphart, 1975).

Similarly, Bryant observed that "In a climate of economic and cultural expansiveness, the traditional social controls of the *zuilen* were no longer tolerable;" also, he believed that Catholics had become sufficiently self-assured, so that they ceased to feel a strong need for the protection that their organizations could offer (Bryant, 1981, p. 61). Shetter, too, contended that strong party-voter ties "largely disappeared," and that secularism and political radicalism ate away at the old foundations of the *zuilens'* authority (Shetter, 1987, pp. 163, 183). So widely acknowledged is this process of depillarization that it has acquired its own name, *ontzuiling*.

And yet one question that this study of euthanasia seeks to raise is whether notwithstanding the decline of *verzuiling*, the culture of decision making that it created and embodied has continued. It is not simply that many leisure and other organizations retain public identities in *zuilen* terms. Nor even that the *zuilen* plainly retain vitality in the public and religious educational systems and in the broadcast media. More fundamentally, there is the issue as to whether the uncanny spirit of compromise and collaboration among elites that constituted the politics of accommodation has persisted in the face of the indisputable decline of the elites themselves.

As to euthanasia, it is a term deriving from the Greek words for good (*eu*) death (*thanatos*), and is of ancient lineage. Today, it has taken on two forms: passive euthanasia, which refers to the withdrawing of extraordinary medical means used to prolong life; and active euthanasia, which refers to actions taken that are intended to cause and do in fact result in death. A bitter and long-standing controversy divides those who consider the two forms of euthanasia morally equivalent (e.g., Kuhse, 1984; British Medical Assoc., 1988; Devettere, 1990; Pellegrino, 1992) from those who consider them different (e.g., Flick, 1991; Brock, 1992; Brody, 1993). But in both types of euthanasia, the purpose is to spare the patient pointless suffering and abridge the agony of injury or disease, and in Saint Thomas More's words, "to free [the patient] from this bitter life" (More, 1963, 4: 186). What has elevated euthanasia to the level where it "is today the most intensively discussed and divisive topic in medical ethics" (Pellegrino, 1991, p. 3118) is the confluence of two developments, one old and one new. As the former director of the Hastings Center put it:

> The combined power of a quasi-religious tradition of respect for individual life and

> a secular tradition of relentless medical progress, creates a bias toward aggressive, often unremitting treatment that appears unstoppable (Callahan, 1989, p. 4).

Euthanasia is one means of stopping the apparently unstoppable.

Without pausing to debate the ethics of euthanasia (see, e.g., Rigter, *et al.*, 1991; Feinberg, 1991; Dworkin, 1993), a few points can be sketched briefly. Nearly every serious Western moral thinker has placed a high value on the preservation of human life. Indeed, it has long been a commonplace to speak of life as "sacred" or "priceless," and to take for granted that those arguing for the permissibility of killing should carry a heavy burden of proof (Kadish, 1987, pp. 109–132). "Thou shalt not kill" may not be absolute, but it is the rule; exceptions have to be justified. Yet life does not exist *in vacuo* nor is it the only good. Sometimes, we care more about avoiding suffering or asserting our autonomy. "Life" is not a mantra, whose repeated utterance dissolves all problems (cf., Goodrich, 1969; Bayliss, 1982). "Quality of life," however, is not a mantra, either (cf., Kamisar, 1958; Kass, 1989). Indeed, the vagueness and banality of the phrase threaten to leave it with only the residual meaning of a life not worthy of being lived. One need not note the German term, *lebensunwertes Leben* (see Burleigh, 1995), to experience some unease.

II. A PRECURSOR: THE ABORTION ISSUE

A number of Dutch observers believe that the nation's experience with euthanasia was influenced or at least foreshadowed by its experience with abortion (e.g., Driesse, *et al.*, 1986, chap. 6). The Dutch penal code of 1881 prohibited abortions. While the law on its face admitted of no exceptions, however, it was generally understood that abortion could be induced to save the mother's life. In fact, a 1911 effort by members of the Parliament (*Staten-Generaal*) to pass a law to this effect was brushed aside by the minister of justice on the ground that it was simply unnecessary; though the Dutch legal system generally assigns little importance to custom (Kisch, 1978, p. 14), the minister found such "medical exceptions" customary, and required no formal authorization. In 1949, a court in Amsterdam broadened the exception to cover psychological-medical indications.

A few years later, other factors – a decline in church-going (Irwin, 1980, pp. 210–212), looser views on sexuality, permissive abortion legislation in England (which not only offered Dutch women a place for their abortions but also helped dispel the stigma of the procedure), and probably most important the *baas in eigen bruik* (boss in your own belly) women's movement – dramatically heightened the salience of the abortion issue. The Dutch government, a coalition of liberal and religious parties, sought to delegate the issue to the medical profession. But the Royal Netherlands Association for the Practice of Medicine (KNMG), the principal medical society, was so divided that it could produce only a set of procedural rules. Even these rules, however, so offended some physicians that conservatives formed the Dutch Physicians Confederation (NAV) and liberals created the Foundation for the Medically Responsible Interruption of Pregnancy (*Stimezo*). Meanwhile, radical lawyers and feminists, some seeing abortion restrictions as unwarranted infringements on personal morality and others as efforts to control women, demanded complete repeal of the law. The doctors' monopoly over the issue was broken.

By the mid 1960s, it was becoming obvious that the legal prohibition against abortion was increasingly ignored in practice; an estimated 17,000–20,000 abortions were being performed annually (Treffers, 1965) as abortion clinics began to spring up. Far from being clandestine, these clinics received significant media attention, usually sympathetic to the plight of the pregnant women. On only one occasion did government prosecutors close a clinic – and in this instance, there was overwhelming evidence of gross mismanagment.

The government, however, still hoped to recapture the abortion issue in a medical bottle. A government appointed commission was created, and in 1971 a bill was proposed that would allow medical teams to authorize abortions under certain specified circumstances. But it was far too late for such a limited approach, and the bill went nowhere. A new government in 1972 made abortion a free issue, leaving it to the *Staten-Generaal* to solve, but by now pro-choice groups were confronted by right to life groups, drawn mainly from orthodox Protestants, and legislation was impossible. When in 1976 the Minster of Justice twice tried to close a clinic performing second trimester abortions, it seemed that maintaining the status quo might be impossible, too. Following an election in 1977, the cabinet reasserted control at the expense of interest groups and the *Staten-Generaal*, returning to the "'classical' politics of accommodation" (Outshourn, 1986, p. 24). Public opinion had come overwhelmingly to favor liberalization, and the abortion law was in tatters.

In 1981, the *Staten-Generaal*, "having given due consideration to . . . the development of opinion with regard to the termination of pregnancy," revised the penal code to permit abortion in cases of "inescapable emergency" – a justification whose apparent narrowness was completely undermined by its extremely liberal practical implementation. In advocating legalization in terms of patient autonomy and social pluralism and flouting the prohibition with devastating brazenness, the defenders of abortion rights not only won a great victory; they also provided an example for other causes.

III. EUTHANASIA: THE BEGINNINGS

Euthanasia did not begin with all of abortion's advantages. There was no foreign example, like the English or later the American, to point to; there was no potent social force, like the women's movement, to make the cause its own; and its chief beneficiaries would not be available to sing its praises and do much of the work that successful legislative efforts required. Like abortion, euthanasia was clearly and unambiguously prohibited by the Dutch penal code. Article 293 provides: "He who robs another of life at his express and serious wish, is punished with a prison sentence of at most twelve years or a fine of the fifth category" (i.e., up to a maximum of 10,000 guilders); and Article 294 provides: "He who deliberately incites another to suicide, assists him therein or provides him with the means, is punished, if the suicide follows, with a prison sentence of at most three years or a fine of the fourth category" (i.e., up to a maximum of 25,000 guilders). These crimes are less serious than murder, for they are not seen as directed against the lives of specific persons; the ostensible victims, after all, have literally requested their own victimization. Still, they are very serious crimes, for they are taken to be directed at nothing less than the sanctity of human life itself.

Yet it must be said that euthanasia found the damp green flatness of Holland a fertile field. The openness of Dutch society to ideas and practices considered progressive had by the 1960s become proverbial. Indeed, many foreign observers grumbled that progressive values had taken hold of the Dutch mind like a new disease gripping a remote island people. There was resistance, of course, and some of the acquiescence was reluctant and fatalistic, but more than any other nation in Europe, Holland was energized by those pursuing the new, hungering for it and reveling in it and the bolder the better. Euthanasia, which to much of the West was not new but on the contrary tied to discredited social Darwinist or eugenic rhetoric, seemed new to the Dutch. And the very fact that they had been spared euthanasia propaganda earlier made them more sympathetic when finally it appeared (van der Sluis, 1989, p. 458) in a booklet by a Dutch professor of neurology, J. H. van den Berg, that was published in 1969. Noting with pitiful examples that medicine had acquired the dubious capacity to sustain pointless life, van den Berg argued for a new ethic obliging doctors to terminate life, either by halting treatment or by actively killing the patient; consent from neither the patient nor his family, van den Berg believed, should be required van den Berg, 1969). Within a year, the booklet went through ten printings, and was soon lauded by prominent Protestant clergymen (e.g., Krop, 1971; *Euthanasia*, 1972). The issue had gained visibility.

The next year, the secretary of state for social affairs and health asked the chairman of the Health Council (*Gezondheidsraad*), an official body that advises the government, the medical profession, and health care institutions, for guidance on the issue. A commission representing various points of view was duly established. In its 1973 report, it concluded that the fundamental distinction was between active and passive euthanasia, not voluntary and involuntary euthanasia; that active euthanasia is and ought to remain forbidden; and that passive euthanasia should become permissible in certain circumstances (N.J.B., 1974). Although falling far short of a complete endorsement of euthanasia, the report added an official element to the practice's legitimacy – and in rejecting the voluntary-involuntary dichotomy and approving involuntary passive euthanasia in limited circumstances, the report marked out a potentially rather radical path.

That same year, 1973, Dutch courts began to carve out a place for euthanasia in the law. In a system governed by criminal and civil codes, there is, of course, something jarringly incongruous about such judicial activism. For the working assumption behind code centered systems is that the codifiers had so thoroughly anticipated contingencies that little more than a mechanical application of rules to specific cases is required of judges. Thus, not only does the Dutch constitution (article 131, *Grondwet*) virtually bar courts from invalidating parliamentary statutes, but settled custom prohibits even dissenting judicial opinions, on the ground that they might encourage belief that the codes are vague or ambiguous.

Yet for some time it had been obvious to many Dutch jurists and lawyers that codes by themselves could not provide clear answers to every question. Indeed, as early as 1919, the supreme court (*Hoge Raad*) asserted its power to interpret the codes aggressively, when in a tort case it issued a ruling that clearly conflicted with the literal wording of the pertinent statute (N.J., 1919). In numerous subsequent cases, Dutch courts repeated this practice, in effect generating layers of precedents that rest awkwardly upon the ostensibly self-sufficient codes. But far from provoking cynicism or discomfort, the practice is nearly always simply reflexively accepted or praised as a commonsensical approach that is fair, workable, and typically Dutch.

The 1973 euthanasia case arose in a district court (*Arrondissementsrechtbank*) in Leeuwarden. Dutch criminal cases are decided not by juries but by judges, usually a panel of three judges, and follow the inquisitorial model, in which judges call and question witnesses and otherwise dominate the proceedings. The Leeuwarden case involved a general practitioner, Geertruida Postma, whose 78 year old mother had been left partially paralyzed and deaf, and had then been moved to a nursing home, where she had been tied to a chair with a restraining belt to prevent her sliding to the floor. "When I watched my mother, a human wreck hanging in that chair," Dr. Postma recalled, "I couldn't stand it anymore" (Parachini, 1987). The mother repeatedly pleaded with the head nurse, "Isn't there anything to kill me, I want to die, I just want to die." Dr. Postma injected her with a lethal dose of morphine.

Dr. Postma was found guilty under Article 293, but sentenced only to a week in prison – and that sentence was suspended, pending a year's probation. The three judges rejected her plea that "rational suicide" assisted by a physician is not prohibited by the law or that the doctor's own psychic distress constituted a defense. But after eighteen physicians signed an open letter to the minister of justice stating that they had practiced active euthanasia – Dutch judges often "adopt statements of scholars" (Kisch, 1978, p. 15) – the court observed that "the average physician in the Netherlands no longer considers it right that the life of a patient be stretched to the bitter end."

More importantly, the court announced a set of criteria under which active euthanasia would be permitted: the disease must be incurable, the patient's suffering unbearable, the patient must freely request his own death, and the patient's physician must perform the euthanasia or participate through consultation (N.J., 1973). The Health Council's advice was superseded.[1]

Notwithstanding its ambivalent ruling, the case elicited substantial support for euthanasia. Millions were moved by Dr. Postma's heartbreaking story and reassured by the court's criteria, particularly its rejection of involuntary euthanasia. Even an opponent of euthanasia conceded that the court was "hailed as very wise, humane, and compassionate" (van der Sluis, 1989, p. 458).

Later in 1973, the KNMG issued a report declaring that "legally euthanasia should remain a crime but that if a physician, after having considered all aspects of the case, shortens the life of a patient who is incurably ill and in the process of dying, the court will have to decide if there was a conflict of duties which could justify the act of the physician" (Gevers, 1987, p. 158).

IV. EUTHANASIA: THE 1980s AND AFTER

Although the KNMG report generated considerable public discussion, a court did not significantly address euthanasia again until 1981, when a Rotterdam district court convicted a layman of assisting a suicide; no physician had even been consulted. Central to the decision was the absence of competent medical personnel able to make professional judgments. Yet the court went beyond this, enumerating nine criteria that must be satisfied if euthanasia were to escape punishment (Leenen, 1987, p. 200): 1) the patient must experience unbearable suffering; 2) the patient must have, when conscious, expressed a desire to die; 3) the patient's request for euthanasia must be voluntary; 4) the patient must have been advised of alternatives and given time to consider them; 5) there must be no other reasonable solutions to the patient's problem; 6) the patient's death must not inflict unnecessary pain on others; 7) more than one person must be involved in the decision; 8) only a physician may actually perform the euthanasia; and 9) great care must be taken in reaching the decision. Defining key terms – "unbearable suffering," "reasonable solutions," and so forth – was left to the medical profession.

The Rotterdam decision generated considerable discussion, most of it positive, and the ministry of justice moved to address euthanasia's unusual legal and moral ambiguities. The ministry's response, however, was wholly procedural and not at all substantive: instead of the normal practice of allowing individual prosecutors to determine whether to proceed, the heads of prosecution at the courts of appeals level were directed to confer and reach a consensus. Yet even by 1981 and 1982, euthanasia remained so unusual that Dutch investigators found that the term was almost never used by physicians and nurses in hospitals nor requested by patients themselves (Hilhorst *et al.*, 1983).

It was in 1982, however, that the most important Dutch euthanasia case began its long journey through the courts. It concerned Maria Barendregt and a general practitioner and frequent euthanasia advocate, P. L. Schoonheim. Mrs. Barendregt, a forceful woman who prided herself on her independent functioning, had moved to a nursing home at age 89 as a result of her declining health. By age 94, she had deteriorated to the point where she was unable to sit or care for herself. Then she fractured her hip, suffered hearing and vision loss, at times was unable to speak, and could neither eat nor drink. Her mind remained alert, however, until eventually she lost consciousness. When she reawakened, she declared that she did not want to repeat the experience, and requested euthanasia. She discussed her condition with Schoonheim, his assistant, and her son, and announced that she wanted "to die as soon as possible." Schoonheim gave her a series of injections that killed her, and informed the medical examiner that he had euthanized her.

A district court at Alkmaar acquitted Schoonheim, stressing Mrs. Barendregt's right to autonomy, the care and seriousness with which the physician had proceeded, and the extensive documentation of her request (N.J., 1983). The prosecutor appealed the decision, arguing that it departed from the standards set down in the Rotterdam case. An Amsterdam court of appeals (*Gerechtshof*) ruled for the prosecution. Given the language of Article 293, the court found the absence of an "adequately crystallized general view on the basis of which euthanasia . . . can be regarded as generally and socially desirable." But although Schoonheim was convicted, he was not ordered punished.

With the help of the Netherlands Society for Voluntary Euthanasia (NVVE), which had been established in 1973 in the wake of the Leeuwarden case, Schoonheim appealed to the supreme court, which in 1984 announced its decision. The supreme court agreed with the prosecution that the claim that Schoonheim was merely assisting Mrs. Berendregt carry out her right of self-determination did not excuse his violation of Article 293. At the same time, though, the supreme court found the court of appeals' ruling fatally flawed: it was much too narrow in failing to consider other excuses (e.g., whether there were other means of eliminating the patient's suffering apart from euthanasia; whether the physician faced a "conflict of duties" between his legal obligation not to kill and his medical obligation to do what good medical practice demands; or "whether, according to prevailing standards of medical ethics, an emergency existed," which could be met only by euthanasia). The supreme court dispatched the case to the court of appeals in the Hague, ordering it to consider the case "from an objective medical perspective" (N.J., 1985).

In the meanwhile, three developments took place. One involved the State Commission on Euthanasia, which had been created by the ministers of justice and health and environmental protection in 1982. The commission consisted of seven attorneys, three physicians, two psychiatrists, two theologians, and a nurse. It was asked by the ministers to set down specific recommendations for active euthanasia, and in 1985 all but two members signed a report concluding that euthanasia should not be punishable by law under certain conditions (when euthanasia is performed by a physician within a setting of careful practice, when the patient is advised of the serious and irreversible nature of his condition and initiates the request for euthanasia, and when a second physician is consulted). The commission also recommended that "no medical worker should be obliged to participate in active termination of life." The commission urged that the government act quickly before the issue was taken up by political parties, and warned that letting the courts construct official policy would be a very slow process (*Euthanasia*, 1985).

Far from settling the controversy, the report antagonized both sides. Advocates of euthanasia criticized it for failing fully to acknowledge the patient's autonomous right to end his life and to comprehend how demeaning prolonged dying can be. Opponents of euthanasia, on the other hand, castigated the report for apparently approving active euthanasia of patients in a persistent vegetative state on the family's request – the report is not clear on this point – and for failing to confine the practice to terminal cases.

A second development concerned the KNMG. There had already been indications that elite medical opinion was generally supportive of euthanasia; for example, the entire May 28, 1983 issue of a leading medical journal, *Nederlands Tijdschrift voor Geneeskunde*, consisted of an editorial defending the practice and a series of articles dealing with such practical problems as physician nightmares and avoiding prosecutors' inquiries. By 1984, the KNMG, fearful that legal uncertainties might permit physician prosecutions, asked the minister of justice to seek a change in the law to allow euthanasia. A substantial number of physicians resigned from the organization in protest, and when the bill was duly proposed, the Council of State (*Raad van State*), a senior advisory body consisting of members of the royal family and certain elder statesmen, took the unusual step of intervening and advising the cabinet to reject the matter. The Council cited the ongoing Schoonheim litigation and the pending State Commission on Euthanasia report.

In response, the KNMG issued a report of its own. Though ostensibly taking no position on the morality of euthanasia, the report declared that it had "based [its] position on the fact that euthanasia is already applied." The report then moved to set down morality-based criteria that physicians practicing euthanasia were urged to follow. In order to protect the patient against coercion or his own transient impulses, the report emphasized that the patient must freely request euthanasia after having thoroughly considered it and that his decision must be stable over time. And in order to guard against physician lapses, the report stated that the physician must be confident that the patient's suffering was unbearable and that a colleague must be consulted before action was taken (KNMG, 1984).

The report left several obvious questions unasked. What of incompetent patients, for example, who are incapable of requesting or consenting? ought the patient's condition to be irreversible? must it be terminal, in the sense that death is clearly in sight?

A third development was the proposal of a bill in the *Tweede Kamer* (or Second Chamber or lower house) of the *Staten-Generaal* by a member of *Demokraten* '66 (Democrats '66), a small, energetic left-wing party with support among young intellectuals. The Wessel-Tuinstra bill – named after its sponsor – would have legalized euthanasia, so that "it would not be punishable in the framework of a careful rendering of assistance to a person who is in a hopeless situation." The Council of State again intervened on the ground that the State Commission on Euthanasia's report had not yet been issued.

Normally, legislation originates in the cabinet, where it is discussed and agreed to before being sent to the *Staten-Generaal*. And so following the issuance of the Euthanasia Commission report, the government announced that it considered legislation on euthanasia unwise, and preferred to permit courts to evolve pertinent criteria. If the *Staten-Generaal* were determined to act, however, the government would offer a proposal. When as a consequence of extraordinary legislative interest this determination became obvious, the government introduced a bill encompassing the Commission's recommendations, except that it required that death be imminent and that the patient be competent and consenting or, if incompetent, that documentation of a prior consent be presented. This provoked a second Wessel-Tuinstra bill, which would have permitted euthanasia for patients in a "hopeless situation" but not necessarily a terminal one. For the third time, the Council of State intervened, blocking voting on the bills in the name of awaiting further rulings from the judiciary.

It was in this context of heightened interest and uncertainty that the court of appeals in the Hague was instructed to rule on the Schoonheim case. And at this juncture, the prosecutor asked the KNMG for its views as to whether a requisite consensus on euthanasia existed within the medical profession. Notwithstanding the fact that many physicians had left the organization over its call for legalized euthanasia and that the KNMG had failed to poll its members on the issue, the KNMG issued a report stating that when a "conflict of duties" like that faced by Schoonheim arose, a doctor might feel compelled to practice euthanasia. Under these circumstances, while euthanasia remains a crime, the physician should be legally without blame.

Relying on the KNMG report, the Hague court found "no norms of medical ethics that forbade his [Schoonheim's] actions," but, on the contrary, numerous physicians who "consider active life termination as medically-ethically possible." On the basis of this declaration, Schoonheim's obvious care in reaching his decision, the "medical insight" that Mrs. Berendragt's death was inevitable, and Mrs. Berendregt's heavily documented desire for euthanasia, the court upheld the acquittal.

The ruling is interesting for the way the Hague court turned the supreme court's instructions on their head. Instead of holding that a medical consensus supported euthanasia, it satisfied itself with noting that there was no medical consensus against it; and instead of finding "objective medical criteria" to justify euthanasia, it settled merely for "reasonable medical insight."

Shortly after the Hague decision, the government asked the Health Council for guidance concerning the carefulness requirements. In March, 1987 the Council responded with six standards: 1) the physician must advise the patient of his condition and prognosis and the possible therapies that may be pursued; 2) after mature reflection and without external pressure, the patient must repeatedly request the doctor to end his life – and both doctor and patient must be convinced that the request comes from the patient; 3) point two is waived if the patient had previously freely written a request that his life be ended, provided that there exists no other solution to his problem; 4) the physician must consult with a colleague; 5) the physician must write a report setting down in detail his compliance with the guidelines, and he must retain the report for five years; and 6) unless patients under age 16 raise a valid objection, their parents must be involved in decisions to end their life (Committee of the Health Council of the Netherlands, 1987).

At about the same time, the KNMG and the Interest Association of Nurses and Nursing Aides jointly published guidelines for cooperation and job demarcation among physicians, nurses, and nursing aides in the carrying out of euthanasia. By dealing with such practical matters of implementation, the Health Council, the KNMG, and the nurses' organization implicitly gave euthanasia itself their imprimatur.

At the end of 1987, the government proposed two bills. One would leave the penal code unchanged regarding euthanasia, except that the maximum sentence would have been shortened from twelve to four and a half years. The other bill provided that "Without prejudice to his responsibilities under the Penal Code, the medical doctor who wishes to follow the explicit serious wish of a patient to terminate the life of that patient should abide by a number of requirements of careful medical behavior," such as consulting with another doctor and with the family. These two incompatible bills left it open to question whether the government intended that physicians practicing euthanasia be liable to prosecution. Neither bill, in any case, was adopted.

In 1993, the government offered legislation that while retaining the prohibition against killing patients or aiding suicides, condensed the various judicial holdings into six rules that if met, would safeguard physicians from prosecution. Specifically, the bill required that the euthanasia request be made "entirely of the patient's free will" and not from outside pressure; that the request be explicit and be made over a period of time; that the patient be adequately informed and capable of weighing the choices; that the patient must be experiencing unacceptable suffering and have no hope of recovery; that the physician consult at least one colleague, who has experience with euthanasia and who agrees with the decision; and that the physician provide a full report to the coroner (*lijkschouwer*) and certify that the rules have been followed.

At first glance, a distant observer might conclude that the legislation was prompted by an urge to bring order to the unruly collection of criteria that had accreted over the years. Certainly, solemn declarations to the contrary notwithstanding (see, e.g., Rigter, 1988, p. 1593; ten Have, 1989, p. 41), pre-existing policy rules had been anything but clear and consistent. Some criteria, for example, recurred in major judicial rulings (e.g., the patient must freely request his own death, euthanasia must be performed by a physician), while others were mentioned once and then dropped (e.g., the patient's disease must be incurable, the patient's death must not inflict unnecessary pain on others). Similarly, the State Commission on Euthanasia, the KNMG, and the Health Council each issued guidelines that differed in certain respects from the other.

Furthermore, judicial decisions and official reports had been replete with terms of boundless vagueness, like "great care" and "objective medical perspective," and ignored such obvious concrete questions as whether suffering encompassed emotional as well as physical pain and discomfort or how euthanasia was to be accomplished.

Yet, in fact, few interested persons seemed preoccupied with or even aware of the congeries of inconsistencies and vaguenesses. On the contrary, the euthanasia legislation debate was very largely a spectacle of moralistic jousting. The only mundane argument to attract attention was the government's contention that the bill would improve public scrutiny of euthanasia by inducing physicians to report cases of it.

Rather, what motivated the supporters of legislation was the recognition that they had steadily gained in strength and could now codify the disparate rulings into a single law and by so doing further legitimate euthanasia. Public support had grown from about two-fifths of the adult population in 1966 to two-thirds in 1986 (Rigter *et al.*, 1991, p. 1593) to about four-fifths in 1988 (Battin, 1991, p. 300) and 1993 (Simons, 1993a).[2]

The struggle was brief. The liberal and socialist parties and physicians groups favored euthanasia; Catholic bishops, several small Protestant sects, and the right-to-life movement opposed it; and the Christian Democratic party, which dominated the center-left governing coalition, wanted only an incremental change in the status quo. The incremental change embodied in the bill carried in the lower house, 91 to 45. "Today the Netherlands abolished the Hippocratic oath," declared one angry right-to-life physician. "I regret to say that we are becoming a barbaric nation" (Gunning, in Simons, 1993b).

V. EUTHANASIA: THE EXTENT OF THE PHENOMENON

How widespread, then, is euthanasia in the Netherlands? It is impossible to answer the question with confidence, for it is universally acknowledged that to avoid possible prosecution prior to the 1993 legislation, physicians rarely reported cases of euthanasia to the ministry of justice, coroner, or other authorities (e.g., Sluyters, 1989, p. 41; Schoonheim, 1991, p. 110; Horgan, 1991, p. 20; Kuhse, 1987, p. 395; Battin, 1991, p. 300; Pollard, 1991, p. 614). As the secretary-general of the KNMG observed, "Most doctors do euthanasia without reporting it" (Brant, 1988, pp. 19–20). This is accomplished simply by pretending that the death was from natural causes, and issuing a death certificate to that effect. "Cause of death," as a government health inspector conceded, "is easy to hide" (van der Wijmen, in Simons, 1993b). That prosecutors seem almost as reluctant to prosecute as physicians do to report – in 1990 of 454 cases of euthanasia reported to the ministry of justice, none was prosecuted (Battin, 1991, p. 300), for example, and in 1987 of 125 cases reported, two were prosecuted (Schoonheim, 1989, p. 111) – appears not to affect physician underreporting to any great extent. Only about 11 cases per year even generate prosecutor inquiries (Pollard, 1991, p. 614).

As to the actual number of patients euthanized, estimates vary. One writer suggests between 2,000–10,000 patients per year (de Wachter, 1989, p. 3316); another 3,000 (Horgan, 1991, p. 20); another 3,000–4,000 (Battin, 1991, p. 300); another at least 5,000 (Schoonheim, 1989, p. 110); others 5,000–8,000 (Angell, 1991, p. 1349; ten Have, 1989, p. 41; Pence, 1988, p. 139); others 5,000–10,000 (Kinsella, *et al.*, 1989 p. 7); another 6,000–10,000 (Appleyard, 1987); another 5,000–20,000 (Fenigsen, 1989, p. 23); and yet others 20,000 (Dessaur and Rutenfrans, 1986). A survey commissioned by the government-appointed Committee to Investigate the Medical Practice Concerning Euthanasia (usually known as the Remmelink Committee, taking the name of its chairman, John Remmelink, attorney general at the supreme court) estimated that in 1990 Dutch physicians performed about 3,735 acts of euthanasia, assisted suicide, and other involuntary life termination; withheld or withdrew treatment that would probably have prolonged life in about 22,538 cases; and treated about 22,538 patients with opioids to alleviate pain in such dosages that life may have been shortened. Put differently, active euthanasia comprised about 2.9 percent of the nation's total deaths, passive euthanasia about 17.5 percent, and the alleviation of pain another 17.5 percent; omitting the 30 percent of deaths where there were no opportunities for medical intervention (e.g., fatal automobile crashes), the percentages of the total deaths were 4.1, 25, and 25, respectively – in the aggregate, 54.1 percent. These figures do not include newborns, children, psychiatric patients, or AIDS patients; their inclusion would doubtless have raised the number (van der Maas, *et al.*, 1991, p. 670).

As for physicians, two points may be made. The first is that most physicians have practiced active euthanasia. Thus, the Remmelink Committee survey disclosed that 54 percent of all Dutch physicians (and, specifically, 62 percent of general practitioners) had performed euthanasia. The second point is that nearly all doctors approve or at least concede the legitimacy of euthanasia. The same survey found that in addition to the 54 percent who had actually performed euthanasia, another 34 percent would be willing to do so under certain conditions, and eight percent would not perform euthanasia themselves but would refer their patients to a physician who would; only four percent of Dutch physicians would never perform euthanasia or make a referral (van der Maas, *et al.*, 1991, p. 671). (Ironically, nursing home physicians, who would seem to have the greatest opportunity to euthanize, were by far the least likely ever to have done so; only 12 percent had ever performed euthanasia. At the same time, though, they were the most likely to say that they would be willing to perform it under certain conditions and the least likely to say that they would never perform it or make a referral.) So widespread is physician support for euthanasia that in 1991 a doctor who had agreed to give his patient a lethal dose but instead had merely given him ordinary tranquilizers was formally reprimanded by the KNMG for failing to keep his promise (Levin, *et al.*, 1992).[3]

How does Dutch physician support for euthanasia compare with that found elsewhere in Western Europe? It is not always easy to say. A working party of the British Medical Association, for example, urged that euthanasia remain a crime (British Medical Association, 1989), and many individual physicians have expressed their agreement (e.g., Norris, 1988). On the other hand, Britain's Institute of Medical Ethics' working party found medically assisted death justified in certain circumstances (Institute of Medical Ethics Working Party, 1990), and 30 percent of British general practitioners accept the concept of euthanasia ('The Attitude of GPs to Voluntary Euthanasia,' 1987). More broadly, a survey of members of the European Society of Intensive Care Medicine suggested that Dutch intensive care unit doctors were only slightly more likely to practice active euthanasia (though they were much more likely to practice passive euthanasia) than physicians in other Western countries (Vincent, 1990, p. 259).[4]

As might be expected, euthanized patients typically suffered from delayed degenerative diseases, most commonly cancer, and were usually over age 65. This is in keeping with general trends in the developed world, where the incidence of sudden, acute afflictions has been replaced by late-arriving diseases that lead to predictable and prolonged declines (see Olshansky and Ault, 1986).

VI. EUTHANASIA: THE POLICY ANALYZED

In reviewing the Dutch policy on euthanasia, the observer is struck by several points. The first is that though the judicial rulings and ultimate legislation were typically justified in terms of patient autonomy, the actual practice has rather openly departed from this norm. On the one hand, patients must not merely request death but must persuade physicians of the rightness of the decision, and physicians often assert their authority to say no. "It may take ten conversations or more before I am convinced," declared one doctor (Cohen, in Simons, 1993b); "You have to ask it again and again," said another ('Choosing Death,' 1993). According to the government-commissioned survey, twice as many patients who requested euthanasia were denied as accepted (van der Maas, *et al.*, 1991, p. 672). This is in line with a physician survey of 1985–86 that reported that only 37 percent of euthanasia requests had been granted (van Wijmen, 1989).

This paternalistic chord is in harmony with the general theme of the judicial rulings and 1993 legislation, which seems designed less to protect the patient from his physician than from himself. For instead of relying upon simple patient consent to trigger euthanasia, the lawmakers have introduced a number of additional requirements in an effort to ensure that this momentous and irreversible decision is not made by mistake. "The issue," as a Harvard law professor declared in a very different context, "is false consciousness" (Kennedy, 1982, p. 572). The false consciousness may be attributable to treatable depression, medical ignorance, fear of becoming a family burden, or sheer impulsiveness. But whatever its etiology, the lawmakers have concluded that the patient is in need of protection against his own proneness to error.

Of course, few notions are as treacherous as false consciousness, which a cynic might operationalize simply as patient decisions of which judges and legislators disapprove. The lawmakers, after all, may be wrong, either in evaluating the patient's decision as a means to an end or in understanding the end that the patient has selected. More importantly, we are accustomed to think that ordinary adults are entitled to make decisions about their own lives, even if the decisions are wrong, and we shrink from the assertion that patients are incapable of making the only decisions that they truly care about. As one patient with amyotrophic lateral sclerosis put it, "If I cannot give consent to my own death, whose body is this? Who owns my life?" (see Rodriguez, in 'Whose Life is it Anyway?' 1994).

Yet if it is easy to rebut paternalism with autonomy, it is hard to avoid concluding that autonomy, like paternalism (Halper, 1980), is a double-edged sword. For ordinary adults, autonomy may rank among the highest values. Intrusions on independence may seem insults designed to displace one's very self. But patients contemplating euthanasia in the context of terrible suffering are hardly ordinary adults. Vulnerable and insecure, they may view autonomy as a cruel hoax or a transparent rationalization for their abandonment. Shame or guilt may induce them to demand autonomy anyway. Or they may call for it as a despairing protest against their perceived neglect and a prayer for better care.[5] Although patients requesting euthanasia may not be exhibiting false consciousness, what they say – however dramatically, emphatically, and incessantly – may not be what they mean. Paradoxically, responding to patients' stated wishes may frustrate their deeper needs (Fletcher, 1987).

Or to put it differently, legislating a right to euthanasia may mean that some patients who do not really want to die feel compelled to do so. For with the possibility of choosing death always explicitly before them they may feel that they must defend their refusal to die in the face of the burden this refusal may impose on family and friends. It is not only that others may pressure a patient to choose euthanasia; it is also that the patient may begin to pressure himself to make that choice. A legal right to die, therefore, may be regarded as an official invitation to die, an invitation of such force that the traditional unspoken assumption that life be continued may be undermined by doubts that eventually cause it to come crashing down (Vellerman, 1992).

At the same time, however, assurance that one can choose suicide may dampen the urge to take one's life by inducing one to put off the decision. Instead of grabbing the opportunity from a fear that it will not recur, one would feel secure to let the uncertain future unfold, reasoning, "I need not rush to end my life because the option will remain." And perhaps as the future unfolds, the anxieties one felt are shown to be unfulfilled. By delaying the decision, suicide may be avoided.

Acceptance of the notion that the severely ill uniquely are entitled to select the timing and means of their own death also underlines their categorization as "other," different, set apart from the rest of humanity. They are not simply potential candidates for death – as are we all – but also potential and legitimate candidate for killing themselves.

Yet if euthanasia in the Netherlands is relentlessly justified in terms of patient autonomy, as a practical matter it often proceeds in the total absence of any patient consent at all. In deaths resulting from treating pain and symptoms with medications that threatened life, doctors discussed the subject with patients in only about 40 percent of the cases; in deaths resulting from withdrawing treatments that would probably have prolonged life, the patient was consulted in only 30 percent of the cases (van der Maas, *et al.*, 1991, p. 672). A 1990 government-commissioned survey estimated that over a thousand patients per year were euthanatized without their explicit and persistent request (van der Maas, *et al.*, 1991, p. 670). It is not hard to imagine that inquiries into the actual deaths would produce numbers that far exceed this figure, which is based solely on admissions freely made by physicians themselves long after the events (on abuses generally, see Gomez, 1991). Certainly, the response of the health ministry spokesman on medical ethics was not reassuring: "It saves the patient that last bitter end" (Visser, in Chapman, 1995).

Still, the frequency of involuntary euthanasia was by far the most commonly heard criticism of the Dutch practice. Indeed, so common was it that in 1995 the KNMG recommended that, whenever possible, doctors ensure that patients give themselves the fatal drug. As a member of the KNMG ethics commission expressed it, "This step is first a decision of the patient, and eating or drinking the substance is a symbol of the patient's responsibility" (Simons, 1995).

Second, the presumption of patient competence – or at least, that the patient had competently issued an advance directive in the past – leaves unaddressed the issue of the incompetent patient who never chose. It appears that in practice physicians make substitute judgments on the basis of their understanding of the patient's best interests. But without formal guidance, "best interest" is a concept of amoeboid contours. Beyond an unspoken resolve to let physicians follow their consciences in the light of their experience and skills, there can hardly be said to be a euthanasia policy for this numerous class of patients at all.

Other questions receive answers in law that bear little relation to practice. To what extent, for example, is end-of-life decision making authority perishable or capable of being given by proxies? In the United States, concern is sometimes expressed that informed consent expressed at one moment should not control at another, when the person is incapable of making his wishes known (Buchanan, 1981, p. 386); or contrariwise, that one should be protected from changing his mind at a point of weakness, comparable to Odysseus' encounter with the Sirens (Schelling, 1989, chap. 6). In the Netherlands, the judicial and legislative insistence on active, informed consent is construed to obviate this problem, despite the fact that euthanasia often occurs in the absence of any consent whatever.

To what extent should the physician's advantages in knowledge, skills, and experience limit the patient's interests in autonomy, privacy, and dignity? In the United States, the physician's medical expertise is ordinarily interpreted as carrying with it a certain moral authority, and he often takes an active role in seeking to shape the patient's or his family's desires (Fletcher, 1987). In the Netherlands, on the other hand, so great is the rhetorical emphasis upon the patient's role that physicians claim that custom forbids them even to raise the subject – a claim that is utterly impossible to take seriously. Magnifying the difficulty is the lens of beneficence, for proponents of euthanasia never miss an opportunity to characterize medicine as the preeminent caring profession and euthanasia as the quintessential act of caring. One Dutch critic characterized this as "smug" and "sanctimonious," another as "flatulent self-congratulation." More troubling, caring suggests a personal relationship, not a livelihood or a bunch of incompatible bundles of impersonal rules. And in common speech, caring seems a corollary of the Golden Rule, though it is doubtful that a healthy person can adequately conjure up the life of the terminally ill and suffering patient, not to say the persistently vegetative one (if, indeed, there is in this case even a self to conjure). It is not idle solipsism to observe that one can never fully know another's best interests; one is inescapably a captive of his own perspective and experiences. And the more "different" the patient – the more seriously ill or disabled – the more difficult it is for another person to put himself in his place, even if that other person is an experienced physician (Ullmann and Pearlman, 1991; Greene, *et al.*,

1989; Zweibel and Cassel, 1989; Ouslander, *et al.*, 1989; Seckler, *et al.*, 1991; Ullmann, *et al.*, 1988). Nor is the possibility of simple abuse to be disregarded. For, although euthanasia physicians may at first be beneficent, how does one "prevent the reasons for merciful killing from being hijacked by the merciless" (Lamb, 1988, p. 62) or by those who accept it merely as standard operating procedure?[6]

Finally, what is even more striking than the flimsiness of the euthanasia criteria is the way they actually operate. Take, for example, the requirement that a physician consult with a second doctor before consenting to perform euthanasia. This requirement is effectively defanged by the question-begging proviso that consultation is necessary only if there is medical doubt as to diagnosis and prognosis. In practice, frequently there is no consultation at all, and when consultation does take place, it is often no more than a brief conversation between colleagues that proceeds without the second doctor even seeing the patient or the medical opinion being put to paper (van der Wal, *et al.*, 1991).

Similarly, the requirement that physicians report cases of euthanasia they have performed is widely flouted. Criticism of this practice led the KNMG in 1995 to emphasize the importance of second opinions and to recommend that the doctor consulted may not have a professional or family relationship with either patient or doctor. Though the number reported almost tripled from 591 in 1991 to 1436 in 1994, even the larger figure is grossly inaccurate. And when reports are filed, they normally produce only perfunctory discussions with indifferent medical examiners. The justice ministry practice allowing euthanasia prosecution only when all the heads of prosecution at the courts of appeal level have consented – itself a departure from standard Dutch prosecutorial procedure – is merely an additional official deterrent to official action. Thus, an observer reading three Dutch defenders of the euthanasia policy announcing that "most physicians performing euthanasia were not prosecuted" (van Delden, 1993, p. 25) can only marvel at their gift for understatement.

VII. EUTHANASIA: THE POLITICS OF ACCOMMODATION?

What is interesting to the observer is that though it is widely known that the euthanasia rules are often vague and inconsistent, nearly always unenforced, and frequently violated, little has been done about it. Until 1993, no legislation was adopted, and even the law was as notable for what it ignored as what it accomplished. What makes all this appear even odder is the Dutch popular reputation as "an earnest, cautious, principled people who take their lives, their attitudes and their religion seriously" (Huggett, 1971, p. 83). In fact, the Dutch themselves seem prone to speak of themselves as "prudent, bourgeois, parochial" (de Jong, 1966–67, p. 5) or "honest, dull, thrifty, self-satisfied" (van Schagen, 1964–65), and their most eminent historian declared:

> Whether we fly high or low, we Dutchman are all bourgeois – lawyer and poet, baron and labourer alike. Our national culture is bourgeois in every sense you can legitimately attach to that word. The bourgeois conception of life is shared by all classes or groups of our people – urban and rural, property owning or not (Huizinga, 1968, p. 112).

When one envisages a people so described, the quality of law abidingness comes quickly to mind.

Yet each society hides behind its presentation to the world, and it would be a mistake to accept the Dutch stereotype at face value. Instead, the euthanasia case seems to be a textbook perfect illustration of Lijphart's politics of accommodation, notwithstanding Lijphart's own belief that the days of *verzuiling* are done (Lijphart, 1971; Lijphart, 1975). It is true, of course, that the interlocking directorates among political parties and familiar pressure groups have faded away. The political role of religion, apart from Protestant fundamentalism, has also declined, and the cohesiveness of organizational networks is not what it was. Yet the politics of accommodation still retains its vitality. The accommodation on euthanasia involves not the grand Catholic, Calvinist, liberal, and socialist blocs, but much more modest interests like the KNMG and NVVE. The old society anchored to four large pillars has evolved into a more pluralistic one of smaller factions and more transient allegiances.

But in the face of these major changes, what remains essentially intact is the set of norms and attitudes that undergirded Lijphart's old politics of accommodation: the eagerness of government to facilitate elite bargaining to settle disputes, the readiness of leaders to compromise without rancor or bitterness, and the willingness of the public to acquiesce quietly in the arrangement. If there is today, as many observers have maintained, less of the characteristic Dutch civility, reserve, politeness, and horror at making a fuss, there is still much of the venerable determination to preserve humane toleration by informally assigning elites the task of working out conflicts among themselves. Thus, while the cast of accommodationists has grown exponentially since Lijphart announced his theory three decades ago, the actual politics of accommodation has not changed much at all.

Superficially, it may appear that accommodation on an issue like euthanasia would be unusually difficult to achieve. There is the apparent either/or choice of life and death, the moral passions of proponents of each point of view, and the concrete memories or plausible anticipations of large numbers of persons. Accommodation here hardly resembles splitting the difference on a public works project or logrolling on an omnibus subsidy arrangement.

All this is not to rule out accommodation on euthanasia, however, but merely to indicate that it may take a somewhat different form; for accommodations arise not from a rigid process but instead from the specific circumstances and character of the dispute. One key element in the euthanasia dispute has been that mainstream elites have perceived no inducement fully to resolve the issue. While zealots contend pro and con, the government and its more centrist opponents have recognized the political and moral ambiguity of the issue. Human dignity and autonomy would seem, at the very minimum, to permit a patient to try to persuade his doctor to cut short a prolonged and painful dying; this is normally reduced to the unanswerable question, whose life is it, anyway? At the same time, this libertarian emphasis may invite abuse – patients may be led to feel an obligation to cease being a burden to others, or may succumb to a transitory depression and choose to die when they might have survived, or may be put to death without their knowledge or consent; physicians may compromise their role as healers by acting as executioners; and religious persons may believe that "people are the creation of God and do not have the right to determine about their own living, about ending it" (Schinkalshoak, in Clines, 1986). It is a complicated dispute, involving ethical principles, religious beliefs, and abiding personal fears – and forthright statements are certain to antagonize segments of the populace. No wonder politicians not motivated by moral certitude would prefer to avoid it until a durable and clear cut consensus eased the burden of decision. And even then, far from reflecting the full force of the consensus, the legislation left inconsistent provisions of the criminal code in place and did not concern itself with the major issue of the incompetent patient.

A second key element is that the profoundly ethical character of the dispute makes it a good candidate for resolution by actors who appear more objective and given to deep thought than mere politicians – like judges and doctors. Thus, repeatedly, the mainstream governing elites – whether in the government or the Council of State – blocked political action on the ground that the issue was better suited to judicial determination, and the courts essentially remanded the issue to practicing physicians. Even the 1993 legislation can be viewed less as restraining the doctor than as protecting him, and the KNMG ethics spokesman justified its recommendation that patients euthanize themselves by observing, "Many doctors find euthanasia an extremely burdensom action, and the patient's participation diminishes the burden slightly" (Dillmann, in Simons, 1993b). All this rests on the presumption that euthanasia is, if not wholly a medical matter, then at least one on which doctors can speak with overpowering authority. It is true, of course, that euthanasia is confined to sick persons under physicians' care, that diagnosis, prognosis, and pain management are major euthanasia concerns and are wholly within the doctors' sphere of competence, and that euthanasia itself is carried out only by doctors. It, therefore, appears to strike most Dutch as natural that the euthanasia decision be considered a medical judgment based upon clinical expertise. This, however, is not to say that medical judgments are, like theorems in Euclidean geometry, free from temporal dross. For one thing, it is not always possible to be certain about prognoses, even involving patients in a persistent vegetative state (Levin, *et al.*, 1992), and this uncertainty feeds disagreement. For another, different physicians may draw different inferences from the same case: one, perhaps more optimistic in temperament or more confident in his ability to manage pain, may view a patient's last weeks as a valuable time; another, with a different orientation, may see the period as a nightmare to be avoided.

Heightening the uncertainty in cases of euthanasia is the fact that physicians often lack a great deal of experience with the procedure. This makes it particularly difficult to decide whether to euthanize patients, who seem to fall in the gray area between obvious acceptance or rejection. In such cases, physicians naturally proceed analogically, comparing the patient before them with similar patients they have treated, observed, or otherwise learned about through the literature or from colleagues. But doctors know that the patient before them is not exactly like the patient with whom he is compared, and that in the real world "identical" is merely an analytical construct that really means "similar." When physicians make relatively few such comparisons, each decision necessarily becomes more speculative, for a plethora of similar and thus reassuring experiences is absent.

By this point, it is plain that the physician is not merely a medical actor, though as a medical doctor that is his title and that is his training. In addition, the physician is a political actor, which is to say that he must not only make "clinical judgments" and practice "good medicine", but also has a role in the authoritative allocation of values in society (Easton, 1953, p. 130). If a relatively few people are affected by his decisions, the impact on these few will be difficult to exaggerate, for nothing less than life and death is involved. The point is not that physicians are inappropriate for this role, for apart from the patient and his family, they may well be better equipped to participate in euthanasia decisions than other available actors. Nor is it that doctors can utilize their authority to act politically through such organizations as the KNMG. The point instead is that doctoring itself can be a political act. It is hard to see, therefore, how clinical rhetoric can transmute euthanasia as a question of public policy into a medical decision. For the expertise of physicians does not extend to ethical questions, even if they arise in medical contexts (Halper, 1989, chap. 3).

However, it was convenient for Dutch politicians and judges to believe or pretend that this was so, for it spared them the political and psychic costs of deciding. Even after the 1993 legislation, the physician's autonomy is far more impressive than the patient's. Certainly, doctors publicly agonize about euthanasia a great deal. "It's a very emotional and wrenching experience," one complained (Leenen, in Simons, 1993a); "I find it very emotional, very taxing," said another (Cohen, in Simons, 1993b); it is "very emotional [and] leaves a scar on the soul," declared a third (Smallhoot, in 'Choosing Death,' 1993). The restraints on physician behavior, however, seem almost entirely internal or psychological. External institutionalized restraints – a hallmark of accountability in democracies – are no more vigorous than a dying patient. In Lijphart's terms, the euthanasia issue has been "depoliticized" (Lijphart, 1971, pp. 129–130).[7]

Why, however, should physicians accept their role? One answer is that it accords with the goal of self-regulation that professions have always sought (and by so doing, effectively safeguards physicians against official prosecution). Another is that it confirms a social judgment as to doctors' high-mindedness and impartiality. Thus, in a very loose and informal sense, the key political, judicial, and medical actors can be said to have colluded in support of the fiction that the desirability of euthanasia is essentially a medical decision. Hence, the KNMG expressed no reservations, for example, about doctors being instructed to govern their performance of euthanasia by recourse to medical "perspective" or "insight,' in spite of the fact that euthanasia had been unlawful and thus legally outside medical perspective or insight. Nor did the KNMG pause to emphasize problems deriving from the physician's complex set of circumstances. For the doctor's task need not be confined to that of an observer, recording events and responding to repeated patient requests; he may also help to form the requests, either through indirect or subtle suggestion, more frontal techniques, or merely providing the patient with information about his condition. Neither court guidelines nor legislation addressed this nor has organized medicine spoken up to counter the simplistic fiction of the passive physician.

The medicalization of euthanasia – and a bland, mechanical medicalization it is, too – is nowhere explicitly acknowledged. This is not a trivial matter, for it helps to ensure that the vast majority of patients and their families – and even many physicians themselves – remain blind to the moral and discretionary nature of the situation. Thus blind, they mistake moral judgments for medical ones, and are more likely to acquiesce. Appearing intelligible, reasonable, and just, the medically cloaked rationale for euthanasia can make a strong case for legitimacy. Medical science is enshrined as a source of moral authority.

What, however, does the Dutch accommodation offer to the opponents of euthanasia? Knowing that they are clearly the weaker side, they do not expect victory. Restraining their adversaries or playing for time is all they can really hope for. Viewed from this very limited perspective, the accommodations – first, the judicial decisions and then, the 1993 law – can be viewed as the best of their poor set of choices. For though effectively condoning euthanasia, they at least withhold unqualified formal legal approval. The legitimacy of the practice thus remains in some small doubt, and the possibility of rolling back events is kept theoretically alive.

VIII. SOME CONCLUSIONS

For generations, it seemed a settled fact that euthanasia was prohibited in the Netherlands. Beginning in the late 1960s, however, a number of developments, social and technological, induced physicians and others to challenge this prohibition. Stressing physician expertise and patient autonomy, their arguments quickly won a significant public following, and by the early 1970s, started to be translated into judicial rulings. Though these rulings did not agree on all particulars, they were united in their contention that courts could create a lawful role for euthanasia. Thus, while lists of legal prerequisites varied from case to case in a somewhat unpredictable and bewildering manner, the main theme remained the same: subject to certain shifting safeguards, the physician was to be free to perform euthanasia. Political leaders accepted a compromise, featuring the old restrictive law, the new liberalized practice, and court decisions that pretended that harmony reigned between the two. But if politicians avoided a potentially nasty divisiveness by transferring the issue to the judiciary, the courts also eluded the problem by passing matters onto physicians. The physicians' discretion, in turn, was legitimated by numerous legal restraints – and because none of the restraints was enforced with any regularity whatever, they were little more confining than common moral admonitions. In the process, a political decision on an ethical issue was transformed into complex technical questions hidden from public view behind the arcana of legal and medical terminologies, traditions, and mystiques.

In such a game, nearly all the actors won. That is why accommodation is so prevalent and so popular. The political leaders sidestepped a costly conflict. The courts' moral authority was reinforced, while they, too, were spared many tough decisions. And the doctors saw their prestige and independence enhanced and the threat of official retaliation all but eliminated. The mass public, for its part, appeared to accept the resolution without complaint. They do not expect politicians to act as moral potentates.

Who lost? Activists pro and con euthanasia were denied legislation unambiguously embodying their principles; each was forced to settle for less than it would like, especially the opponents of euthanasia. Patients, who enjoy virtually none of the protection so sonorously proclaimed in the penal code and in judicial rulings, also lost; the prevalence of involuntary euthanasia means that those who might have chosen life are instead put to death. But the passive vulnerability that argued for these protections also ensured that the patients and their families are unlikely to make a commotion.

By 1993, however, the accomodationist consensus had persisted for so many years that it no longer appeared particularly radical or disruptive to recognize the general understanding in law. Indeed, to its sponsors, it may well have seemed an overdue formality not worth much attention. It is essential to recall, however, that the law formalized only prior judicial rulings, not settled practice. That settled practice had for years departed from these rulings – and would, presumably, continue to depart from the new law – was barely acknowledged by the law's proponents. Thus, despite the new law and the new candor it was said to embody, the old patterns proceeded, indifferent to legislation and as hardy as ever.

Department of Political Science
City University of New York
Graduate School and Baruch College
New York, New York, U.S.A.

NOTES

[1] The court, which landed Dr. Postma for her "perfect purity of motives," may well have been actuated by sympathy for the defendant, and the applause the ruling won suggests that this sympathy was widespread throughout the nation. It was not noticed that The court, which lauded Dr. Postma for her "perfect purity of motives", may well have been actuated by sympathy for the defendant, and the applause the Dr. Postma failed to meet the court's criteria, which presumably had been framed with her in mind. Since the mother was not receiving anti-depressants, it could not have been determined either whether her suffering was unbearable or her request to die untainted by reversible clinical depression; nor was Dr. Postma her mother's attending physician nor had she acted after a consultation with him. Dr. Postma's suspended sentence, on the other hand, is quite in keeping with the well known mildness of the Dutch criminal justice system. Two years earlier, for example, only 21 percent of those convicted of homicide were sentenced to more than three years imprisonment, only five percent of those convicted of armed robbery, and only one percent of those convicted of rape (Hulsman, 1978, p. 295).

[2] In the United States, public opinion has shifted, so that overwhelming majorities favor permitting a permanently comatose patient's family to decide whether to end life support (Coyle, 1990, p. 36) and allowing a terminally ill patient to refuse life-sustaining treatment (Richards and Cohen, 1990). Majorities also support Dr. Jack Kovorkian well-publicized suicide assistances (Beck, 1990; Malcolm, 1990), and *Final Exit*, a how-to suicide manual written by the president of the Hemlock Society, Derek Humphrey (1991), was one of the best-selling books of the decade (Ames, 1991).

[3] Ignoring right-to-die advance directions have led to lawsuits in the United States (Levin, 1968).

[4] Support for euthanasia among American physicians is hard to estimate, though episodic evidence suggests a significant sympathetic minority. An anonymous letter to the *Journal of the American Medical Association* described a doctor's administering a lethal dosage of morphine to a dying patient undergoing great suffering; the doctor was exhausted,

inexperienced, and unacquainted with the patient, and did not even trouble to examine her chart or consult her regular physician; the patient's suicidal request was so cryptic as to be indecipherable; nonetheless, of the 150 physicians who wrote to *JAMA*, about a fifth expressed their approval ('It's Over Debbie,' 1988; Lundberg, 1988).

5 This is illustrated by the famous American case of Elizabeth Bouvia (*Bouvia v. County of Los Angeles*, 1987; *Bouvia v. Superior Court*, 1986). Suffering from severe cerebral palsy, feeling abandoned by her husband and parents, she persuaded a California court to honor her wish to starve herself to death – winning national notoriety in the process. At this point, she changed her mind, and decided not to die. Also widely publicized was Larry McAfee, a quadriplegic who won court permission to turn off his ventilator, decided to live, and saw his story told in a 1993 television movie, "The Switch" ('McAfee, Larry, 39', 1990).

6 In a well known case, a Dutch physician aided in the suicide of a woman who was neither old nor ill but who, depressed over the death of her sons, had tried to kill herself and had convinced the doctor that she would try again in a "violent," "horrid" fashion. The doctor was acquitted (Kaufman, 1994; a similar story is told in Segers, 1987).

7 Interestingly, though euthanasia in the United States is often considered in a context of highly costly patient care (Leibniz and Prihoda, 1984; Callahan, 1988), euthanasia in the Netherlands is rarely even discussed in economic terms – and on those rare occasions when the subject is raised, it is ordinarily brushed aside with obvious impatience (e.g., Rigter, 1988, p. 32), although the Dutch are far from indifferent to the cost of health care.

BIBLIOGRAPHY

Althusius, J.: 1932 [1603], *Politica Methodice Digesta*, Harvard University Press, Cambridge.

Ames, K.: 1991, 'Last Rights', *Newsweek* (August 26), 40.

Angell, M.: 1989, 'Euthanasia', *New England Journal of Medicine* 319, 1348–1350.

Appleyard, B.: 1987, 'The Last Appointment', *London Sunday Times* (June 7), 18–22.

Battin, M. P.: 1991, 'Euthanasia: The Way We Do It, The Way They Do It', *Journal of Pain and Symptom Management* 6, 298–305.

Bayliss, R.: 1982, 'Thou Shalt Not Strive Officiously', *British Medical Journal* 285, 1373–1375.

Beck, M.: 1990, 'The Doctor's Suicide Van', *Newsweek* (June 18), 46–47.

Bouvia v. County of Los Angeles: 1987, 195 Cal. App. 3d 1075, 241 Cal. Reptr. 239.

Bouvia v. Superior Court: 1986, 179 Cal App. 3d 1127, 225 Cal. Reptr. 297.

Brandt, M. I.: 1988, 'Legal Status of Practice Remains Cloudy in Holland', *New South Wales Doctor* 20 (September), 19–20.

British Medical Association: 1988, *Euthanasia: Report of the Working Party on Euthanasia*, British Medical Association, London.

Brock, D. W.: 1992, 'Voluntary Active Euthanasia', *Hastings Center Report* 11 (March-April), 10–22.

Brody, H.: 1993, 'Causing, Intending, and Assisting Death', *Journal of Clinical Ethics* 4, 112–117.

Bryant, C. G. A.: 1981, 'Depillarization in the Netherlands', *British Journal of Sociology* 32, 56–74.

Buchanan, A. E.: 1981, 'The Limits of Proxy Decisionmaking for Incompetents', *UCLA Law Review* 29, 386–408.

Burleigh, M.: 1995, *Death and Deliverance: 'Euthanasia' in Germany 1900–1945*, Cambridge University Press, Cambridge.

Callahan, D.: 1989, 'Can We Return Death to Disease?', *Hastings Center Report* 19 (January-February; special supplement), 4–6.

Callahan, D.: 1988, *Setting Limits: Medical Goals in an Aging Society*, Simon and Schuster, New York.

Caralis, P. V. and Hammond, J. S.: 1992, 'Attitudes of Medical Students, Housestaff, and Faculty Physicians toward Euthanasia and Termination of Life-sustaining Treatment', *Critical Care Medicine* 20, 683–690.

Chapman, S.: 1995, 'Dead Wrong', *Chicago Tribune* November 12, I, 17.

'Choosing Death: *A Health Quarterly/Frontline* Special', 1993, Public Broadcasting System, March 23, 9–11 p.m. (EST).

Christakis, N. A. and Asch, D. A.: 1995, 'Characteristics Associated with Decisions to Withdraw Life Support', *Journal of the American Medical Association* 273, 367–372.

Clines, F. X.: 1986, 'Dutch are Quietly Taking the Lead in Euthanasia', *New York Times*, October 31, A4.

Committee of the Health Council of the Netherlands: 1987, *Zorgvuldigheidseisen Euthanasie* [Requiring Carefulness in Euthanasia], U 1887/RvZ/ir. Den Haag, Gezondheidsraad, March 26.

Coser, L. A.: 1956, *The Functions of Social Conflict*, Free Press, Glencoe.

Coyle, M.: 1990, 'How Americans View High Court', *National Law Journal* (February 26), 1.

Daalder, H.: 1974, 'The Consociational Democracy Theme', *World Politics* 26, 604–621.

Dahrendorf, R.: 1959, *Class and Class Conflict in Industrial Society*, Stanford University Press, Stanford.

De Jong, O. J.: 1966–1967, 'Dutch Protestantism', *Delta* 9 (Winter), 5–16.

De Wachter, M. A. M.: 1989, 'Active Euthanasia in the Netherlands', *Journal of the American Medical Association* 262, 3316–3319.

Dessaur, C. J. and Rutenfrans, C. J. C.: 1986, *Mag de doktor Doden?* [May the Doctor Kill?], Querido, Amsterdam.

Devettere, R. J.: 1990, 'The Imprecise Language of Euthanasia and Causing Death', *Journal of Clinical Ethics* 1, 268–274.

Driesse, M. H. N., van der Kolk, H., van Nunen-Forger, W. A., and de Marees van Swinderen, E.: 1986, *Op Leven en Dood* [A Matter of Life and Death], Block end Zonen, Dieren.

Dworkin, R.: 1993, *Life's Dominion: An Argument about Abortion, Euthanasia and Individual Freedom*, Knopf, New York.

Easton, D.: 1953, *The Political System*, Knopf, New York.

Eldersveld, S. J., Koolman, J., and van der Tak, T.: 1981, *Elite Images of Dutch Politics: Accommodation and Conflict*, University of Michigan Press, Ann Arbor.

Euthanasia: 1985, Den Haag, Staatscommissie Euthanasie.

Euthanasia: Zin en Begrensing van het Meddische Handelen. Pastorale Handreiking [Euthanasia: The Meaning and the Limits of Medical Actions. Manual for Ministers]: 1972, Generale Synode der Nederlandse Hervormde Kerk, Den Haag.

Feinberg, J.: 1991, 'Overlooking the Merits of the Individual Case: An Unpromising Approach to the Right to Die', *Ratio Juris* 4, 31–51.

Fenigsen, R.: 1989, 'A Case against Dutch Euthanasia', *Hastings Center Report* 19 (January-February; special supplement), 22–30.

Flick, J.: 1991, 'The Due Process of Dying', *California Law Review* 79, 1121–1167.

Fletcher, J.: 1987, 'The Courts and Euthanasia', *Law, Medicine and Health Care* 15, 23–30.

Gevers, J. K. M.: 1987, 'Legal Developments Concerning Active Euthanasia on Request in The Netherlands', *Bioethics* 1, 156–162.

Gomez, C. F.: 1991, *Regulating Death: Euthanasia and the Case of the Netherlands*, Free Press, New York.

Goodrich, T.: 1969, 'The Morality of Killing', *Philosophy* 44, 127–139.

Greene, M. G., Adelman, R. D., and Charon, R.: 1989, 'Concordance between Physicians and Their Older and Younger Patients in the Primary Care Encounter', *Gerontologist* 29, 808–813.

Halper, T.: 1989, *The Misfortunes of Others: End-Stage Renal Disease in the United Kingdom*, Cambridge University Press, New York.

Halper, T.: 1980, 'The Doubled-Edged Sword', *Milbank Memorial Fund Quarterly* 58, 472–499.

Heilig, S.: 1991, 'Murder or Mercy? The Debate Over Active Euthanasia Has Only Just Begun', *HEC Forum* 3, 95–98.

Hendin, H.: 1994, 'Scared to Death of Dying', *New York Times* (December 16), A39.

Hilherst, H., Kragt, V. and Banders, A.: 1983, Euthanasie in het Ziekenhuis, de "Zachte Dood" bij Ziekenhuis patienten [Euthanasia in the Hospital: The Gentle Death of Hospital Patients], de Tijdstroom, Lochem.

Horgan, J.: 1991, 'Death with Dignity', *Scientific American* 264, 17–20.

Huggett, F. E.: 1971, *The Modern Netherlands*, Pall Mall Press, London.

Huizinga, J. H.: 1968, *Dutch Civilization in the Seventeenth Century and Other Essays*, A. J. Pomerans (trans.) Collins, London.

Hulsman, L. H. C.: 1978, 'The Dutch Criminal Justice System from a Comparative Legal Perspective', in D. C. Fokkema, Chorus, J. M. J., Hondius, E. H., and Lisser, E. C. (eds.), *Introduction to Dutch Law for Foreign Lawyers*, Kluwer, Deventer, pp. 289–377.

Humphrey, D.: 1991, *Final Exit: The Practicalities of Self-Deliverance and Assisted Suicide of the Dying*, Hemlock Society, Eugene, Oregon.

Institute of Medical Ethics Working Party on the Ethics of Prolonging Life and Assisting Death: 1990, 'Assisted Death', *Lancet* 336, 610–613.

Irwin, G. A.: 1988, 'Belgium and the Netherlands', in G. A. Dorfman and P. J. Duigan (eds.), *Politics in Western Europe*, Hoover Institution Press, Stanford, pp. 79–109.

Irwin, G. A.: 1980, 'Patterns of Voting Behaviour in the Netherlands', in R. P. Griffiths (ed.), *The Economics and Politics of the Netherlands Since 1945*, Martinus Nijhoff, The Hague, pp. 199–222.

'It's over Debbie': 1988, *Journal of the American Medical Association* 259, 272.

Kadish, S. F.: 1987, *Blame and Punishment*, Macmillan, New York.

Kamisar, Y.: 1958, 'Euthanasia Legislation: Some Non-Religious Objections', *Minnesota Law Review* 42, 969–1042.

Kass, L. R.: 1989, 'Neither for Love nor Money: Why Doctors Must Not Kill', *Public Interest* 94, 25–46.

Kennedy, D.: 1982, 'Distributive and Paternalist Motives in Contract and Tort Law with Special Reference to Compulsory Terms and Unequal Bargaining Power', *Maryland Law Review* 41, 563–658.

Kinsella, T. D., Singer, P. A., and Siegler, M.: 1989, 'Legalized Active Euthanasia: An Aesculapian Tragedy', *American College of Surgeons Bulletin* 74, 6–9 and 56.

Kisch, I.: 1978, 'Sources of Netherlands Law', in D. C. Fokkema, Chorus, J. M. J., Hondius, E. H., and Lisser, E. C. (eds.), *Introduction to Dutch Law for Foreign Lawyers*, Kluwer, Deventer, pp. 13–16.

KNMG: 1984, 'Standpunt inzake euthanasie' ['Viewpoint Concerning Euthanasia'], *Medische Contact* 39, 990–998.

Knox, R. A.: 1991, 'Dutch Study Reports Doctors Hasten Death in 38 Percent of Cases', *Boston Globe*, September 13.

Kossman, E. H.: 1978, *The Low Countries, 1780–1940*, Clarendon Press, Oxford and New York.

Krop, M. A.: 1971, 'Het Recht op de Dood' ['The Right to Die'], *Medische Contact* 26, 1033.

Kuhse, H.: 1984, 'A Modern Myth: That Letting Die is Not the Intentional Cause of Death – Some Reflections on the Trial and Acquittal of Dr. Leonard Arthur', *Journal of Applied Philosophy* 1, 21–30.

Kuhse, H.: 1987, 'Voluntary Euthanasia in the Netherlands', *Medical Journal of Australia* 147, 394–396.

Lamb, D.: 1988, 'Down the Slippery Slope', in *Arguing in Applied Ethics*, Croom Helm, London.

Leenen, H. J. J.: 1987, 'Euthanasia, Assistance to Suicide, and the Law: Developments in the Netherlands', *Health Policy* 8, 197–206.

Leibniz, J. and Prihoda, R.: 1984, 'The Use and Costs of Medicare Services in the Last Two Years of Life', *Health Care Financing Review* 5, 117–131.

Levin, H. S., Saydjari, c., Eisenberg, H. M., Foulkes, M., Marshall, L. F., Ruff, R. M., Jane, J. A., and Marmarov, A.: 1992, 'Vegetative State After Closed-Head Injury: A Traumatic Coma Data Bank Report', *Archives of Neurology* 48, 580–585.

Levin, T.: 1991, 'Igoring "Right to Die" Direct, Medical Community is Being Sued', *New York Times* (June 2), I, 1.

Lijphart, A.: 1968, *The Politics of Accommodation: Pluralism and Democracy in The Netherlands*, University of California Press, Berkeley.

Lijphart, A.: 1971, 'Verzuiling', in A. Hoogerwerf (ed.), *Verkenningen in de Politiek [Shedding Light on Policy]*, Samson, Alphen aan de Rijn, pp. 26–38.

Lijphart, A.: 1975, *The Politics of Accommodation*, 2 ed., University of California Press, Berkeley.

Lijphart, A.: 1977, *Democracy in Plural Societies: A Comparative Exploration*, Yale University of Press, New Haven.

Lipset, S. M.: 1960, *Political Man*, Doubleday, Garden City.

Lundberg, G.D.: 1988, '"It's over Debbie" and the Euthanasia Debate', *Journal of the American Medical Association* 259, 2142.

Malcolm, A. H.: 1990, 'Giving Death a Hand', *New York Times* (June 9), I, 6.

'McAfee, Larry, 39: Sought Right to Die': 1995, *New York Times* (October 5), D23.

More, T.: 1963, *The Complete Works*, E. Surtz and J. H Hexter (eds.), Yale University Press, New Haven.

N.J. [Nederlands Jurisprudentie]. 1919, No. 161.

N.J. [Nederlands Jurisprudentie]. 1973, No. 18.

N.J. [Nederlands Jurisprudentie]. 1983. No. 407

N.J. [Nederlands Jurisprudentie]. 1985. No. 106.

N.J.B. [Nederlands Juristenblad]. 1974. No. 1254.

Nordlinger, E. A.: 1972, *Conflict Resolution in Divided Societies*, Harvard University Press, Cambridge.

Norris, P.: 1988, 'Euthanasia in The Netherlands', *British Medical Journal* 298, 386.

Oliver, J.: 1990, 'The Moral Consensus on Euthanasia', *British Journal of Hospital Medicine* 44, 11.

Olshansky, S. J. and Ault, A. B.: 1986, 'The Fourth Stage of Epidemiological Transition: The Age of Delayed Degenerative Diseases', *Milbank Memorial Fund Quarterly* 64, 355–391.

Ouslander, J., Tymchuk, A. J., and Rahbar, B.: 1989, 'Health Care Decisions among Elderly Long-Term Care Residents and Their Potential Proxies', *Archives of Internal Medicine* 146, 1367–1372.

Outshourn, J.: 1986, 'The Rules of the Game: Abortion Politics in the Netherlands', in J. Lorendduski and J. Outshourn (eds.), *The New Politics of Abortion*, Sage, London, pp. 5–26.

Parachini, A.: 1987, 'A Dutch Doctor Carries Out a Death Wish', *Los Angeles Times* (July 5), VI, 1.

Pellegrino, E. D.: 1991, 'Ethics', *Journal of the American Medical Association* 265, 3118–3119.

Pellegrino, E. D.: 1992, 'Doctors Must Not Kill', *Journal of Clinical Ethics* 3, 95–102.

Pence, G. E.: 1988, 'Do Not Go Gently into that Good Night: Mercy Killing in Holland', *American Journal of Medicine* 89, 139–141.

Pollard, B. J.: 1991, 'Medical Aspects of Euthanasia', *Medical Journal of Australia* 154, 613–616.

Powell, G. B.: 1970, *Social Fragmentation and Political Social Hostility: An Austrian Case Study*, Stanford University Press, Stanford.

President's Commission for the Study of Ethical Problems in Medicine and Biomedical Research: 1983, *Deciding to Forego Life Sustaining Treatment: Ethical, Medical, and Legal Issues in Treatment Decisions*, Government Printing Office, Washington.

Rachels, J.: 1986, *The End of Life*, Oxford University Press, Oxford.

Richards, C. and Colen, B. D.: 1990, 'Polls: Most Favor "Right to Die" Laws', *Newsday* (June 10), 15.

Rigter, H.: 1988, 'Euthanasia in the Netherlands: Distinguishing Fact from Fiction', *Hastings Center Report* (January/February, special supplement), 19, 31–32.

Rigter, H., *et al.*: 1991, 'Euthanasia Across the North Sea', *British Medical Journal* 297, 1593–1595.

Schelling, T. C.: 1984, *Choice and Consequence*, Harvard University Press, Cambridge.

Schoonheim, P. L.: 1989, 'The Right to Die: The Euthanasia Discussion in the Netherlands', *Journal of Cancer Education* 4, 109–112.

Seckler, A. B., Meier, D. E., Malvihill, M., and Paris, B. E.: 1991, 'Substituted Judgment: How Accurate Are Proxy Predictions?' *Annals of Internal Medicine* 115, 92–98.

Segers, J. H.: 1987, *Ouderen Over Euthansie* [Older People on the Topic of Euthansia], Stichting Schuilplaats, Veenendaal.

Shetter, W. Z.: 1987, *Netherlands in Perspective: The Organization of Society and Environment*, Thornes, London.

Simons, M.: 1993a, 'Dutch Move to Erect Law Making Euthanasia Easier', *New York Times* (February 8), A1.

Simons, M.: 1993b, 'Dutch Parliament Approves Law Permitting Euthanasia', *New York Times* (February 9), A10.

Simons, M.: 1995, 'Dutch Doctors to Lighten Rules on Mercy Killings', *New York Times* (September 11), A3.
Sluyters, B.: 1989, 'Euthanasia in the Netherlands', *Medico-Legal Journal* 57, 34–43.
Spicker, S. F.: 1978, 'The Execution of Euthanasia: The Right of the Dying to a Re-formed Health Care Context', in S. F. Spicker (ed.), *Organism, Medicine, and Metaphysics*, Reidel, Dordrecht, pp. 73–94.
Spicker, S. F.: 1979, 'Do We Have a Right to a Natural Death? Ethical, Legal and Medical Considerations', *Connecticut Medicine* 43, 42–45.
Steiner, J.: 1974, *Amicable Agreement versus Majority Rule: Conflict Resolution in Switzerland*, University of North Carolina Press, Chapel Hill.
Ten Have, H. A. M. J.: 1989, 'Euthanasia in the Netherlands: The Legal Context and the Cases', *HEC Forum* 1, 41–45.
'The Attitude of GPs to Voluntary Euthanasia': 1987, *British Medical Journal* 294, 1294.
Treffers, P.: 1965, *Abortus Provocatus en Anticonceptie* [Induced Abortion and Contraception], Stenford Kroese, Leiden.
Ullmann, R. F. and Pearlman, R. A.: 1991, 'Perceived Quality of Life and Preferences for Life-Sustaining Treatment in Older Adults', *Archives of Internal Medicine* 151, 495–497.
Ullman, R. F., Pearlman, R. A., and Cain, K. C.: 1988, 'Physicians' and Spouses' Predictions of Elderly Patients' Resuscitation Preferences by Physicians and Nurses', *Western Journal of Medicine* 150, 705–707.
Van Delden, T. M., Pijnenborg, L. and Van der Maas, P. J.: 1993, 'The Remmelink Study: Two Years Later', *Hastings Center Report* 23 (November-December), 24–27.
Van der Berg, J. H.: 1969, *Medische Macht en Medische Ethiek* [Medical Power and Medical Ethics], G.F. Callenbach, Nijkerk.
Van der Maas, P. J., Van Delden, J. J., Pijnenborg, L., and Looman, C. W.: 1991, 'Euthanasia and Other Medical Decisions Concerning the End of Life', *Lancet* 338, 669–674.
Van der Sluis, I.: 1989, 'The Practice of Euthanasia in The Netherlands', *Issues in Law and Medicine* 4, 455–465.
Van der Wal, G., van Eijk, Leenen, H. J. J., and Spreeuwenberg: 1991, 'Euthanasie en hulpbij, zelfdoding coor huisartsen' [Euthanasia and assisted suicide by general practitioners], *Medisch Contact* 47, 171–180.
Van Schagen: 1964–1965, 'The Fault of the Dutch', *Delta* 7 (Winter), 17.
Van Schendelen, M. P. C. M.: 1984, 'Consociational Democracy: The Views of Arend Lijphart and Collected Criticisms', *Political Science Reviewer* 15, 143–183.
Van Wijmen, F. C. B.: 1989, *Artsen en het Zelfgekozen* [Physicians and the Self Elected], University of Limbrugh, Maastricht.
Vellerman, J. D.: 1992, 'Against the Right to Die', *Journal of Medicine and Philosophy* 17, 665–681.
Vincent, J. P.: 1990, 'European Attitudes towards Ethical Problems in Intensive Care Medicine: Results of an Ethical Questionnaire', *Intensive Care Medicine* 16, 256–264.
'Whose Life is it Anyway?': 1994, *60 Minutes*, CBS, J. Fages and S. Gartlett (producers), November 20.
Zweibel, N. R. and Cassel, C. K.: 1989, 'Treatment Choices at the End of Life: A Comparison of Decisions by Older Patients and Their Physician-Selected Proxies', *Gerontologist* 19, 615–621.

GEORGE J. AGICH

WHY SHOULD ANYONE LISTEN TO ETHICS CONSULTANTS?*

I. INTRODUCTION

Why should physicians, nurses, patients, or their surrogates/families follow recommendations given by ethics consultants? This question is particularly apt in the context of a *Festschrift* for Stuart Spicker who has been singularly committed to the task of improving the functioning of healthcare ethics committees. Through his Educating Healthcare Ethics Committees Project and the journal, *HEC Forum*, he has sought to cultivate a critical and responsible discourse about the functioning of ethics committees. It is hoped that the following will contribute to this cultivation.

Understanding the nature of the influence exercised in the course of ethics consultation is usually approached by focusing on the unique qualifications or credentials of ethics consultants (Ackerman, 1987; Barnard, 1992; Cranford, 1989; Fletcher and Hoffmann, 1994; Grunfeld, 1990; Jonsen, 1992; LaPuma and Priest, 1992; LaPuma and Schiedermayer, 1990, 1992; Marsh, 1992; Thomasma, 1991; Zaner, 1984). Such an approach, however, prejudices the question, because it assumes that ethics consultants can, under some circumstances and with appropriate "qualifications," provide authoritative advice. Our question, however, concerns not what entitles an individual to give advice authoritatively, but rather how other individuals come to accept or follow the advice given. Phrasing the question in this way is meant to shift the discussion away from the ethics consultant and onto the process of ethics consultation itself in the belief that it is only within the practice of ethics consultation that an acceptable account of the authoritative function of the ethics consultant can be developed.[1]

The ethics consultant's authority is immediately recognized in the giving of advice only because authority is modeled in terms of its most prominent features and in terms of the most obvious kinds of influence that it exerts. For authority to work at all, however, there needs to be a recognition and acceptance of the authority figure or of his utterances that makes possible the authoritative advice giving. This shadowy feature

H. T. Engelhardt, Jr. (ed.), The Philosophy of Medicine, 117–137.

of authority is signaled in this essay's title, "Why Should Anyone Listen to Ethics Consultants?" Listening is different than receiving advice or following advice. I argue that it is a more basic aspect of the ethics consultant's authority than either receiving or following advice. Because of its diffuse nature, however, it is all too easy to overlook it and to underestimate its contribution to the establishment of the ethics consultant's authority.

II. SOCIAL ROLE AUTHORITY OF THE ETHICS CONSULTANT[2]

The expansion of ethics consultative activities by committees, consultative teams, or individual ethics consultants is, no doubt, prompted by many factors including medicine's own aversion to law, the recognition that certain classes of clinical decisions such as withholding and withdrawing life sustaining treatment should occur within the doctor patient relationship rather than under court supervision, as well as the Joint Commission for Accreditation of Healthcare Organizations (1996, pp. 73–100) requiring a process for addressing and resolving ethical problems arising in the care of patients and for protecting patient rights. Ethics consultation activities are one widely used procedure to achieve these ends. These developments attest to the social and historical significance of bioethics as well as the fact that the everyday and leading edge of bioethics involves its fulfilling certain vital practical needs occurring within the clinical practice of medicine. Besides these external factors, ethics consultation is supported by reasons internal to the practice of ethics consultation. The most important of these reasons involves the way that the ethics consultant is accepted by those involved in the case under discussion. Acceptance of the ethics consultant's involvement in a case is a prerequisite for acceptance of the consultant's recommendations. Indeed, it is more basic, because without it the ethics consultant could have no significant involvement. The full sense in which the ethics consultant – as opposed to the consultant's advice – is accepted is an artifact of the consultative process itself and cannot be fully understood without situating it within the context of the particular case. Whatever influence the ethics consultant exercises in the course of consultation depends upon such an emergent process of acceptance.

To understand this process properly, one needs to recognize that it involves a distinctive kind of authority, namely, *social role authority* (Agich, 1995, pp. 276–277). This authority involves two components:

epistemic and competence authority. While epistemic and competence authority are important aspects of the ethics consultant's authority, they need to be contextualized within the practice of ethics consultation. The question posed in this paper, namely, why should anyone listen to an ethics consultant, forces this contextualization by stressing the contribution of other individuals who interact with the ethics consultant in the consultative process. This shift of attention is significant, because it underscores a point critical to understanding fully the authority of ethics consultants, namely, that it is constituted within the practice itself. I use the term *authority in action* or *enacted authority* to identify the kind of authority that is an occurrent feature of the process of ethics consultation.

III. AUTHORITY IN ACTION

Authority in action is constituted and justified through its enactment and not by reference to something else. In this sense, *authority in action* has to be differentiated from being *an authority* and being *in authority*. It is always possible and sensible to ask that someone who or claimed to be *an authority* give reasons to establish that he has the capacity or knowledge establishing him as an authority. It is also sensible to inquire of one who is *in authority* about the specifics of the office or position that the individual holds and about the rules and procedures that establish the authority attached to the position. In contrast, *authority in action* is not dependent on anything external to the role relationship itself. As such, its constitution and justification depends upon a mutual recognition and acceptance by others that is part of the normally unreflective world of everyday life (Schutz, 1967; Schutz and Luckman, 1973). For this reason, the enacted authority of the ethics consultant is best understood in terms of features emergent in the practice of ethics consultation itself and not the office or position that the ethics consultant might hold or in terms of the knowledge or capacity that he might possess. In the setting of a particular clinical case, the way that other individuals perceive the ethics consultant contributes significantly to the consultant's status and influence. In this sense, authority in action is constituted both by the actions and intentions of the individual enacting the role and by other social agents who experience and interact with the authority figure. Although both sides of this relationship are essential and although both sides must be considered in a full account of ethics consultation, this discussion is focused on the contribution that other participants in the ethics consultation process make toward constructing the authority of the ethics consultant.

It might be argued that acceptance of an ethics consultant within a particular case is often based on the perception of health care providers that the consultant possesses certain credentials, expertise, or office. Because of this, our understanding of the *authority in action* of the ethics consultant will ultimately collapse into the consultants being *an authority* or being *in authority*. I do not deny that the expertise or office of the ethics consultant plays some role in garnering acceptance of the consultant in a case. However, I argue below, such an understanding of the acceptance of an ethics consultant is far too narrow, and simply fails to capture the rich array of dimensions and factors which contribute to the acceptance of the ethics consultant and the authority actually exercised by the ethics consultant. Of these factors, the consultant's credentials or office are but a small part. Even though being *an authority* or being *in authority* might influence other health care professionals within an institution, it is unlikely to influence families who are unfamiliar with the ethics consultant or, even, ethics consultation.

The operative notion here is that ethics consultation is a practice within which the role of the ethics consultant is sustained. As a result, the ethics consultant's authority hangs not on institutional or formal characteristics, but on the enactment of the role itself. Saying this does not commit me to the view that ethics consultation is not the product of institutional and political forces, but only that those conditions do not establish the enacted authority of the ethics consultant. The authority of the ethics consultant is a practical authority that must be understood within the context of the practice itself.

IV. PARENT AND POLICE

One way to express the sense in which ethics consultation involves an enacted authority is by comparing it to two paradigms of authority, namely, the police officer and the parent. The parental role illustrates the way that authority is enacted in the practice of parenting itself and the police officer's role serves as a paradigm of being in authority that is institutionally founded and established in terms of formal rules and procedures. The police officer gains his authority in virtue of his office and the rules (laws) that he enforces. The parent, on the other hand, does not have parental authority in virtue of biological fact or legal determination, but rather it is earned in the day to day raising of the child. Thus, one can be (biologically or legally) a parent, yet not be recognized or respected as such in one's relations with the child. Of course, because we speak of parenting as a state as much as a process or practice, it is easy to overlook this important point. Parenting is one of those human activities that is precariously established in the dynamic of a relation with another. In that process, tensions and conflicts routinely arise. Despite the work of child rearing specialists, there is no parental rule book that one can follow reliably. Importantly, parental authority is bound up with the affection and commitment of the parent to the child and the concern of the parent for the child's well-being. As a result of these commitments, a sense of respect emerges in the relationship that grounds the authority that the parent exercises.

Parental authority is thus not so much an authority over the child, but a cooperative authority that is enacted for the child. It is a feature of the relationship, not something that is imposed on one individual by another who has no commitment and no concern for the other. Viewed in these terms, authority is thus a pervasive feature of the parent-child relationship and not something superadded thereto.[3] Institutional structures and social conventions are, of course, important influences on parent-child relationships, because they establish a range of empirical possibilities and they make some possibilities more difficult than others. However, I ignore these influences in the present context of discussion, because they have a marginal impact on the distinctive way that parental authority is constituted in the parent-child relationship.

Ethics consultation is like being a parent in that its authority derives from the actual process of doing an ethics consultation as distinct from institutional or other external characteristics of the consultant. The dynamic authority of the parent is thus a better model for the authority in action demonstrated by ethics consultation. Quite unlike a police officer, an ethics consultant lacks an official position or badge of authority. Even if an ethics consultant were in authority in virtue of the position or office that he held in a particular institution, he would be in authority only over those individuals subject to institutional rules. Individuals not directly subject to institutional rules or procedures and individuals not acquainted with the office or position of the ethics consultant as such could not be expected to defer to the consultant's being in authority.[4] If one is concerned to understand the distinctive character of authority in parent-child relationships, then one must critically study particular parent-child relationships to identify the common places where authority is affirmed, distorted, or rejected. Similarly, to understand the function of authority in ethics consultation, it is essential that one attend to the process of ethics consultation.

As in being a parent, it matters how ethics consultation is enacted. Of course, one might immediately object that how a role is enacted matters across a wide range of roles, some of which fit the in authority model. It matters, for example, how responsibly the police officer acts, because the police officer can abuse his legal authority in ways that violate a citizen's civil rights. The point here, however, is not to argue that ethics consultation is unique among social roles in that its enactment is central, but rather that the authority that emerges in the course of a social practice is fundamentally different from the kind of authority claimed to occur whenever one is in authority in virtue of an office or position. Simply holding the office in question may be a *prima facie* basis for saying that a person is in authority, because the authority in question derives from the office, formal rules and procedures more than from the individual who occupies the office or position. However, in the case of authority in action, authority depends more subtly on an occurrent social acknowledgment or acceptance of the authority figure and on how the consultant acts in the role. Thus, it matters a great deal who occupies the role, a point that is only partly recognized in the debate over which professional is best qualified to do ethics consultations (Ackerman, 1987; Barnard, 1992; Cranford, 1989; Grunfeld, 1990; Jonsen, 1992; LaPuma and Schiedermayer, 1990, 1992; Marsh, 1992; Morreim, 1983; Thomasma, 1991; Zaner, 1984). Focusing on professional qualifications, the debate ignores the personal qualifications or skills that are currently central in establishing the authority of the ethics consultant. Whatever skills and qualifications the individual brings to the role, they are important not by way of credentialing the ethics consultant, but as exhibited in conducting an ethics consultation.

V. INFLUENCE AND IN AUTHORITY CONCERNS

Various individuals interact with the ethics consultant in the course of an ethics consultation. Since these interactions occur over time, in a variety of settings, and often involving a plethora of utterances amenable to a variety of interpretations, the complexity is quite remarkable. For example, an ethics consultant might discuss a case with attending, consulting, and resident physicians, nurses, social workers, respiratory therapists, physical therapists, dietitians, pastoral counselors, as well as the patient and various family members or friends. These interactions might occur over hours, days, or weeks in various settings, such as patient's room, nursing station, hallway, or conference room. These interactions might involve written communication in the medical record, phone conversations, or in person face-to-face discussions with individuals or groups of individuals. In the course of these interactions all of these participants will communicate concerns, problems, likes, dislikes, preferences, aversions; they will express a range of emotional states, beliefs, desires, and thoughts. All of these utterances are amenable to a wide range of interpretations; no wonder then that communication occlusions of various sorts are regularly reported as the main reason ethics consultations are requested and the main preoccupation of ethics consultants. In the course of these interactions, it is understandable that each of these individuals will exercise some influence on one another. Some of this influence will be intentional, others not. Some of the influence will advance the goals of ethics consultation, others impede its progress.

Influence and authority, though related, are not necessarily the same. One can influence another without one's influence having an authoritative basis or without oneself being recognized as an authority. Nonetheless, the interrelationship of influence and authority in ethics consultation reinforce the enacted character of the ethics consultant's action. This interrelationship is apt to be misunderstood whenever authority is understood primarily as in authority. Whenever the ethics consultant's authority is interpreted in terms of the in authority model, people naturally worry about its employment and the ways that institutional or legal rules might legitimate or limit its deployment.[5]

Regarded in terms of in authority, however, ethics consultation is a remarkably pallid example of authority. Ethics consultants function at the pleasure of others. They do not have jurisdiction over problematic cases that are defined in advance, but must be mobilized to action by others involved in the case. They consult and offer advice and recommendations, but not commands (Agich, 1995). Because they are consultants, they require authorization to become involved in a case. As a result, a good deal of their authority is derivative and established within the consultation process itself.

Focusing on the qualities and qualifications of the ethics consultant in terms of in authority ignores the autonomy and moral responsibility of the agents who accept the advice or recommendations. Viewed in term of an in authority model, the question of the moral agency and responsibility of the individuals receiving recommendations from the ethics consultant is sidestepped. The prospect that they might autonomously follow the advice of an ethics consultant without succumbing to force or undue pressure is not seriously considered, because the analysis does not take seriously the practical need for authoritative advice in a wide range of everyday circumstances. This is the central point that needs to be stressed. True authority does not override autonomy, but augments it. This point, however, is apt to be misconstrued whenever authority is viewed exclusively as in authority, because the model of in authority has been taken in modern social science as involving the exercise of force or power over individuals. A central problem that arises in terms of this model is how can the use of power or force be justified. In construing authority in terms of the concept of legitimacy, authority is thought of as power that has been made legitimate. As a result, attention to authority in modern social science has been dominated by a concern with maintaining cohesion in society not simply by force, but a sense of "legitimacy" attached to power. This has lead to the development of the view of authority as legitimate power or, as Max Weber (1947, p. 154) put it in initiating this line of thought, "the legitimate use of physical force."

Regarding authority in terms of in authority, however, the actual practice of ethics consultation and the function of authority therein is obscured.[6] The mention of power or force associated with in authority conjures up the specter of coercion and with it worries about paternalism. It is ironic that even parental relationships can be regarded in what are essentially negative terms when viewed from this perspective. Parentalism, however, is not paternalism either linguistically or in practice. The dominance of the model of the police officer in discussions of authority, however, has obscured some of the everyday meanings of parental authority that need to be recovered if ethics consultation is to be properly understood. Under the sway of in authority, the question becomes less what justifies an individual's accepting the advice or recommendations of an ethics consultant than what justifies the power wielded by the authority figure. However, as Hannah Arendt (1958, p. 82) has rightly observed: "authority precludes the use of external means of coercion; where force is used, authority itself has failed." This observation about authority in general is also particularly apt for ethics consultation.

In this context, it is also helpful to differentiate a *de jure* sense of authority which presupposes some set of legal conventions, system of rules, or method of entitlement by which it might be determined who shall have this particular right and *de facto* sense of authority in which a person is capable of eliciting a distinctive kind of allegiance, belief, or obedience that involves a deference, respect, or trust (Friedman, 1973, p. 125). In this later use, the term authority calls attention to a particular type of influence that one individual has over another which is not as such dependent on rules or other external entitlements, but only on the nature of the relationship itself.

Influence, however, is a slippery term. It suggests an objection, namely, that achieving the effects attributed to influence is really to reduce authority to nothing more than a byproduct of meeting the immediate psychological needs of individuals who accept the advice. As a result, the objection runs, influence is just a hidden form of coercion and all the more questionable for that reason. Although it does not involve physical force, but rather operates through psychological means, it is still objectionable. As a result, speaking of influence is rightly a cause for concern and caution, because the objection seems to reduce the whole process of ethics consultation to a process of eliciting psychological effects without critical ethical analysis and responsible choice by the individuals involved. One aspect of this objection is quite correct, namely, that authority cannot be based only on psychological characteristics, on meeting the psychological needs of individuals alone. Obviously, people might earnestly believe that an individual is an authority and might earnestly follow the authority's advice, yet be wrong in their belief. It is important to note, however, that such an objection could be made for virtually any kind of authority including in authority and not just authority in action. For example, people might and frequently do accept the word of officials who are in authority even when the occupant of the office or position is inept, ignorant, or gained the office by illicit means and people sometimes follow orders of individuals who act as if they were in authority when in fact they are not. Thus, while it is true that considerations such as the strength of one's faith or trust in an authority figure can not provide a sufficient basis for justifying authority, it is important to recognize that being able to recognize someone as an authority does seem to be a necessary prerequisite for authority to exist.

The importance of an authority being accepted by others extends well beyond the realm of health care and ethics consultation. For example, H. L. A. Hart (1961, pp. 92–107) in describing the function of law, expounds at length on the requisite "rules of recognition" which establish the acceptance of law by those subject to it. In so doing, Hart tells us that a legal system which does not achieve a requisite measure of conformity or acceptance is not at all able to function in an authoritative fashion. Because the authority of the ethics consultant, unlike that of the law, is not supplemented by recourse to sanctions, the question of what contributes to the acceptance of the ethics consultant becomes even more vital.

VI. INFLUENCE AND RELIANCE

Authority's need for recognition can be understood by considering other common experiences of everyday life. We routinely rely on others for services, advice, and assistance; in our conversations and interactions we exert influence and are, in turn, influenced by others. Clearly, not all of this influence or reliance on others involves treating the other as an authority, but appreciating the nuances of influence and the common ways that we rely on others can help us to understand the way that the authority in action of the ethics consultant is constituted. For simplicity's sake, I focus on the phenomenon of reliance as a distinctive form of influence that is relevant for understanding ethics consultation.

Reliance is a feature of authority that is common in the everyday world where we naturally and routinely rely on others. In fact, it would be impossible to conceive of a society in which individuals did not exhibit interdependence. To be sure, bioethics' stress on autonomy understood in terms of self-determination encourages the loss of this valuable truth (Agich, 1990b; 1993). Reliance on others involves a fundamental recognition of the other as co-constituting the world about us which is why we share the belief that the world of everyday life is 'objective' rather than purely subjective. Because its meaning is constituted not simply through my own acts of intentionality, but through those of others, objects are experienced as shared and imbued with social meanings. Wherever there is perceptual uncertainty or confusion, for example, at a busy airport when I am not quite sure that the flight number which was announced is my own, I without thinking ask someone whether they heard the flight number distinctly. If they answer affirmatively, I usually rely on their report. If I doubt their report or am especially anxious about not missing my flight, I might ask a ticket agent or airline official. In any event, the reality that I confirm or disconfirm is one that is mediated by the meanings and interpretations of others. It goes without saying, of course, that an enterprise as complex as traveling from one point to another by air involves intricately coordinated actions of multitudes of individuals upon whom I routinely and quite unreflectively depend.

No one can function in a society, much less an advanced western industrialized society, without regularly relying on the actions or advice of others and without being influenced by others. Rather than limiting autonomy, reliance on others can actually expand it. I can decide, for example, to have my furnace replaced, but I may not know how to do so. I am decisionally capable, but executionally inept. A heating contractor might be relied on to install a new furnace. Similarly, I may want the most efficient furnace on the market, but I also want a reasonable pay back period so that I can recover the cost of the furnace during my ownership of the house. A heating consultant/engineer who does not sell or install a particular brand of furnace might be a more reliable source of advice and recommendation. In both instances, my autonomy can be enhanced. In either case, it is reasonable to use the services of another, indeed, to rely on their advice without needing to have them certified as authorities on either the installation of furnaces or their selection. In neither case do I surrender my own responsibility for the decision nor the basic goal of my decision, but have allowed each to advise me regarding how it is that I might actually achieve my goal. Such reliance is an important component of all exercises of authority. Whether I take the heating contractor or the heating consultant/engineer as an authority, I rely on the advice as authoritative. The advice is authoritative in the quite literal sense that it conveys information which I would not otherwise have, information which I believe important for undertaking the action that I desire, namely, purchasing a high efficiency furnace, but which I am not at present in a position to acquire on my own. In effect, I take the advice as authoritative even if I do not regard them as authorities.

The child's reliance on the parent partly involves the parent's superior knowledge, but that only partly explains parental authority. Parents undoubtedly have, to some extent, wider experience and greater knowledge than the child. It is important to note that the qualification, "to some extent," is needed, because it will appear obvious upon any reflection that the parent must cease to be an epistemic authority for the child at certain points of development and in terms of certain subject matters. Nonetheless, one might still insist that the parent is an authority for the child and that this sense of authority sometimes involves advice giving and sometimes not. In the case of explicit advice giving what is at stake is not always whether the advice given is rightly based on knowledge, for example, as when one gives accurate directions, correctly spells a difficult word, or demonstrates a logical proof. The parent is as much, or more so, an authority when he gives other, non-cognitive or value-based advice – advice that we would readily regard as ethical. If such advice involves knowledge, it is not a knowledge of a distinct subject matter such as ethical theory or metaethics, so much as it is a practical knowledge that is expressed in rules or recipes of action, cautionary tales, and aphorisms. It is, to appropriate a term from Michael Polyani, a distinctively *personal knowledge*, because the knowledge includes the commitment and passion of the parent. Recognizing the parent's concern for the child gives the child a basis for accepting parental advice. Accepting parental advice does not mean that the child must simply follow the advice. It means that the child takes the advice to heart, makes it a component of his own decision making, and gives it due weight in setting a course of action. In the end, a child may have many reasons for following or not following a parent's advice. At developmental stages characterized by oppositional behavior, the reason might simply be that a course of action has been endorsed by the parent! Other reasons might include, for example, emotional, experiential, or financial considerations. Any and all of these may be at stake in individual settings. The sense, however, in which a parent is an authority figure for the child goes beyond any specific advice given, but rather involves the inculcation of a wider set of beliefs and values that guide the common purposes that the child shares with the parent. In this sense, a parent is accepted as a *de facto* authority precisely because he is the parent. This fact is a feature of the parent-child relationship, not a feature of any legal or formal institutional framework.

VII. ACCEPTANCE OF THE ETHICS CONSULTANT

The case of ethics consultation is similar. Individuals have many reasons for requesting an ethics consultation; some of these reasons are related to the reasons that support the acceptance of the consultant as an authority, others are not. Sometimes, individuals interact with an ethics consultant without having explicitly requested the ethics consultation themselves; so they might accept an ethics consultant without having thought through their need for the consultant's services. For example, nurses and physicians might have to deal with an ethics consultation requested by a family wanting to discuss termination of care or a family might have to deal with an ethics consultant who has been called to a case by a physician. If, however, the ethics consultant correctly identifies the superficial and latent problems involved in the case, he will have a communicative edge that others involved in the case do not have. It is important to note that this kind of communicative edge always gives an individual a kind of authority, but it is an authority that is different from epistemic expertise. It is also different from in authority. The ethics consultant's authority is built up in the very interactions that make up the ethics consultation. Importantly, establishing credibility is not a one-sided achievement on the part of the ethics consultant; rather, it critically depends on the cooperative constitution by others involved in the case. No matter how reluctant or unenthusiastic their initial reception of the ethics consultant might be, acceptance by others constitutes a *prima facie* basis for the ethics consultant's participation in the case and confers an operative kind of authority on the ethics consultant. This basis may, of course, not meet ethical or legal requirements, for example, when it excludes a competent patient's acceptance or even awareness that an ethics consultation has been called. Nevertheless, such a general acceptance by others involved in the care of the patient is necessary if the ethics consultant is to function authoritatively within the case. Obviously, if a patient is decisionally capable (or, if the patient is not decisionally capable, a surrogate), then his authorization is essential to the enactment of authority by the ethics consultant. Even without it, however, an ethics consultant can be authorized by health professionals to advise or assist them to understand their ethical responsibilities (Agich and Youngner, 1991).

At some point, however, the ethics consultant must receive a stronger authorization to engage in his definition, discussion, and analysis of the problems and ethical issues at stake in the case, but at least a tacit acceptance must occur for the ethics consultant to be able to function in a case at all. For example, an individual physician who is not pleased that an ethics consultation has been called by a family member or nurse, nonetheless, might tacitly accept and so authorize the involvement of the ethics consultant if he does anything short of prohibition. Although reactions short of adamant refusal can, nonetheless, be regarded as a tacit acceptance, acceptance is not an approval that *authorizes* the ethics consultant's involvement. Full authorization occurs only when the significant individuals in a case, that is, individuals who have a standing and a stake in the case, accept the ethics consultant's involvement. Not all individuals have the same standing or stake in a case and, so, a hierarchy is usually present with the patient (or patient's surrogate) and attending physician, in that order, at the top.

Even when the acceptance is very weak, a mere acquiescence or passivity, it provides the ethics consultant with an opening within which to construct a basis for cooperation or, at least, involvement. Engaging in such interactions, particularly when there is some degree of opposition to the involvement of an ethics consultant, is an important test of the consultant's ability to identify shared values and to contribute to a rational discussion of the case. Some kinds of opposition can be construed as forms of acceptance of the ethics consultant's involvement in the case in the basic sense at stake here. These forms of opposition will certainly constitute problems that the consultant must identify and work through in the case, but they are sometimes surmountable. How effectively the consultant does surmount this opposition is itself an expression of the consultant's *authority in action.*

Surmounting opposition can thus be a precondition for the authority in action to develop, because without the acknowledgment and acceptance of significant stakeholders in the case, the ethics consultant would lack an opening for his work.[7] Individual physicians and other health care professionals, more so than patients or their surrogates, will be influenced by the stock of credibility that the ethics consultant might have built over time. The family or patient's acceptance of the ethics consultant – assuming that there is no prior knowledge or acquaintance – will derive from either of two sources. Either it will reflect the health professionals' attitudes about the ethics consultant or it will be constituted on the basis of interactions with the family and patient.

For example, the degree to which a consultant identifies and recognizes the patient's and family's own concerns establishes a basis for accepting the ethics consultant as an authority, particularly in situations in which other health care providers are believed (by the family or patient) to misunderstand their concerns. Sometimes this identification will be aided by other health professionals' characterization of problems and concerns for the ethics consultant. Sometimes, the ethics consultant will by demeanor, style of communication, and empathy establish a direct linkage with the patient or family that provides a basis of trust. It is important to note that acceptance by patient or family is an important source of authority for the ethics consultant. Like the acceptance of attending physicians and other health professionals, it forms the basis for ethical and legal authorization. Legally, it can provide the most basic and fundamental authorization for the ethics consultant. However, it is a mistake to think that this is the only or most important form of authorization (Baker, 1989; Veatch, 1987, 1989a, 1989b). There are clearly other, though more elusive, sources of authority on which the ethics consultant draws. The good will and support of physicians as well as nurses and other health care professionals empowers the ethics consultant to function in ways that achieve practical results. Arranging a family conference, getting consultants to coordinate recommendations and an integrated plan of care, displaying and discussing conflicts and disagreements openly and nonjudgmentally, and mediating or negotiating between opposing parties are all made possible because the ethics consultant is permitted to adopt an authoritative stance or posture by physicians and other health professionals. Like the parent whose rebellious teenager stubbornly and foolheartedly refuses to take even common sense advice, an ethics consultant cannot function as an authority unless his efforts to define and help resolve problems arising in the care of patients are acknowledged and in a general way accepted by the interested parties involved in the case.

VIII. ACCEPTING HELP, LISTENING AND FOLLOWING ADVICE

The advice of the ethics consultant should be accepted on two grounds. First, the situation into which the ethics consultant is called requires that other individuals involved in the case recognize that they need help. The need for help does not so much establish the ethics consultant as an authority as it makes possible the participation of the ethics consultant in the case and makes possible the interaction of the ethics consultant with other involved individuals so that the consultant can bring his analytical and communicative capacities to bear on the case. Making this participation and interaction possible is a necessary, but not sufficient condition for the ethics consultant to function authoritatively. It is necessary not in theoretical terms, but practically. That is, unless others involved in the case recognize or admit that they stand in need of advice, information, or support, the process of ethics consultation will not be instituted. No matter how tentative or conditional the acceptance of the ethics consultant is, it rests on a recognition, hope, or belief that the consultant might be of help. Thus, recognition of a need for help supports the entrance of the ethics consultant in a case.

Besides this necessary condition, another condition is required which affords a sufficient basis for someone's following the advice of an ethics consultant, namely, that the interaction with the ethics consultant provides a reason for acting, believing, or thinking in a certain way. Affording a reason for acting, believing, or thinking, however, may occur either as a direct result of the utterances of the ethics consultant or as a by-product of the ethics consultation. The first alternative is clear enough, but the second needs explication. Even in cases where options are clear and where decision making is uncontested, a need for support might still be present. Reiterating what is already known and validating or supporting a decision already made is an authoritative function of the ethics consultant that can serve to confirm or validate a decision and the reasons supporting it. Opening a moral space (Walker, 1993). for candid discussion of alternatives is an important service that the ethics consultant provides, because it allows the reasons to emerge that make the ethical decisions made acceptable to the parties involved. Thus, the process of listening to an ethics consultant is actually much broader than literally accepting advice. The authoritative involvement of the ethics consultant in a case consists in providing an authoritative permission for a discussion to occur, a question to be asked, or expressing doubts regarding a course of action or decision. Much like an adult child who phones a parent to inform him of an important life decision, individuals involved in ethics consultation seek "approval." This need for approval may be more a subjective than an objective need, but it is a need nonetheless. In fact, this phenomenon of giving approval or permission is quite an important, but overlooked aspect of ethics consultation. It can occur even under the guise of explicitly giving advice whenever the advice given conforms to the beliefs and values of the individuals involved in the case. In those situations, the advice of the ethics consultant confirms their sense of reality, confirms the acceptability of their choices. Approving or giving permission is an important feature of authority in action that is likely to be overlooked in any account of ethics consultation that does not take seriously the practical and situational nature of ethics consultation.

IX. CONCLUSION

Why, then, should one listen to an ethics consultant? *Listening* is a strategically ambiguous term referring both to the initial acceptance or authorization of the ethics consultant's involvement by the significant stakeholders in the case and only secondarily to the acceptance of the advice or recommendations offered. Thus, a first reason for listening to an ethics consultant is that the ethics consultant is called to the case or accepted in the case by the very individuals who receive his advice. As a *consultant* he is authorized and so his words at least have a face validity. This face validity is based on the need for assistance felt or judged by the attending or consulting physicians, nurses, patient, or family/surrogates involved in the case. The face validity established is not a logical or epistemic validity as such, but rather a practical validity that precisely opens a space for the interpretive involvement of the ethics consultant.

A corollary of this point is that other individuals, besides those who request the involvement of the ethics consultant, must recognize that a problem or question exists which needs resolution as well as that they themselves have not been able to satisfactorily resolve the issue at hand. This recognition is extremely important, because it not only opens a space for the ethics consultant, it provides an audience ready or at least somewhat willing to listen to what the consultant has to say. Saying that these other individuals comprise an audience does not imply that they play only a passive role. From what already has been said, it should be clear that they interact with the ethics consultant and so dynamically shape the conversation. The simple acceptance or willingness to participate provides an authorization that is essential for the further development not only of the consultative process, but for the enactment of authority therein. This *authorization by others* sometimes occurs as individuals explicitly consent to the involvement of the ethics consultant, but more often the authorization emerges and is itself a tacit product of the communicative interaction that makes up the ethics consultation. As Carl F. Friedrich has pointed out,

> All authority is in the last analysis the authority of communications. . . . The capacity of human beings to participate effectively in such discourse lends to these communications a quality which enable the communicants to look upon them as "their own," as something which they have taken hold of, as much as "it" took hold of them. The subject of an authority is not a passive but an active member of the community in which such authority prevails. Through that augmentation and implementation which authority provides, the subject becomes engaged (1954, p. 312).

Second, the ethics consultant should be listened to whenever the consultant has earned the respect of the individuals involved in the case. The consultant earns their respect and attention by himself responding to their concerns and questions and enhancing communication about the case. The authority of the ethics consultant thus ultimately relies not on formal rules, procedures, or qualifications of the ethics consultant, but on the action of the ethics consultant in the particular clinical case. Acceptance of the utterances of an ethics consultant often cannot be justified on strictly epistemic grounds, because these grounds can hardly be revealed in any detail in most cases. The ability to give reasons that underlie the epistemic authority of the ethics consultant, however, is nothing if people do not see a need for this knowledge. For the knowledge to be practical, it must be enacted in the course of the consultation; its relevance must be *shown* and not simply said. The authority in action of the ethics consultant is the practical enactment of ethics consultant's role in interaction and communication with the other participants in the case. Like the parent, the ethics consultant's advice or interpretation is heard and taken seriously to a significant extent because a degree of respect is earned that makes the consultant someone whose advice or interpretation should be carefully considered. Although ethics consultation, like all examples of authority, can be regarded as providing reasons for acting, believing, or thinking in a certain way, the reasons proffered in the course of an ethics consultation have to respond to the felt needs of the individuals involved. The sense in which one is inclined to accept or rely on the authority enacted in the course of ethics consultation is thus not adequately explainable or reducible to other kinds of authority, but is best understood as a practical product of the ethics consultative process itself. The ethics consultant thus practically establishes his worthiness to be listened to in the myriad ways that he establishes effective communication with and among others in the case.

Department of Bioethics
The Cleveland Clinic Foundation
Cleveland, Ohio, U.S.A.

NOTES

* This paper benefitted throughout from discussions about authority in ethics consultation with Thomas May, Ph.D. His incisive comments on an earlier draft helped to sharpen the analysis considerably.

1 This move is signaled by the work of thinkers like Richard Flathman (1980) and Bruce Lincoln (1994) who see questions of authority as situated, contextual questions within a social, communicative practice.

2 I use the term ethics consultant to refer to an individual or group of individuals engaged in a concurrent case review in which recommendation or advice is given. I use this term generally and do not differentiate ethics consultation conducted by committees, teams, or individuals, except to the extent that these differences bear on the issue of the authoritative character of the recommendations proffered.

3 Clearly, too many parent-child relations are distorted in the modern world. These distortions, however, do not contradict the analysis offered. Instead, the possibility that the authority relation will be distorted is always present in parent-child relations just as its possibility cannot be logically ruled out in ethics consultation.

4 Even if the ethics consultant were in authority, it seems likely that the ethics consultant's actual exercise of authority would be best unpacked in terms of the ethics consultation process itself. One compelling reason for looking to the process of ethics consultation is the realization that in authority itself can be understood only in terms of a practice of authority in which it is acknowledged and recognized.

5 These are not insignificant concerns; however, they are not sufficiently elastic to adequately capture the actual ways that authority operates in the course of ethics consultation. Hence, they have only marginal interest in the present essay given its primary focus on the operation of authority in the ethics consultative process, not its legitimization. For a framework for thinking about the issues of legitimization, see Agich, 1990a.

6 It also obfuscates other kinds of authority and the social processes by which authority is sustained, but that is another matter. An important approach to these wider questions about the function of authority is found in Lincoln, 1994.

7 Sometimes this acceptance or acquiescence is achieved by reminding the physician of a medical staff or administrative rule that permits and authorizes health professionals and family members to request ethics consultations without the prior approval of the physician. Whether such is a good rule or not is not an issue. What is at issue is whether this institutional rule constitutes an exception to my claim that the ethics consultant's authority is established in action rather than supported by institutional rules or other external considerations. I pointed out that this does not imply that external considerations and institutional rules are irrelevant, but only that they do not define authority in

action of the ethics consultant. All a medical staff rule that permits ethics consultations to occur without the prior approval of the attending physician does is to encourage others involved to request an ethics consultation without having to seek the attending physician's approval. Once a consultation is called, of course, the ethics consultant will need to communicate and interact with the attending physician. Unless such interactions are completely refused, these interactions will involve a minimal kind of acceptance of the ethics consultant's involvement in the case.

BIBLIOGRAPHY

Ackerman, T. F.: 1987, 'The Role of an Ethicist in Health Care', in G. A. Anderson, *et al.* (eds.), *Health Care Ethics*, Aspen Publishing Company, Rockville, Maryland, pp. 308–320.

Agich, G. J.: 1990a, 'Clinical Ethics: A Role Theoretic Look', *Social Science and Medicine* 30, 389–399.

Agich, G. J.: 1990b, 'Reassessing Autonomy in Long-term Care', *Hastings Center Report* 20(6), 12–17.

Agich, G. J.: 1993, *Autonomy in Long-Term Care*, Oxford University Press, New York and Oxford.

Agich, G. J. and Youngner, S. J.: 1991, 'For Experts Only? Access to Hospital Ethics Committees', *Hastings Center Report* 21(5), 17–25.

Arendt, H.: 1958, 'What Was Authority?', in C. J. Friedrich (ed.), *Authority: Nomos I*, Harvard University Press, Cambridge.

Baker, R.: 1989, 'The Skeptical Critique of Clinical Ethics', in B. Hoffmaster, *et al.* (eds.), *Clinical Ethics: Theory and Practice*, Humana Press, Clifton, New Jersey, pp. 27–57.

Barnard, D.: 1992, 'Reflections of a Reluctant Clinical Ethicist: Ethics Consultation and the Collapse of Critical Distance', *Theoretical Medicine* 13, 15–22.

Cranford, R. E.: 1989, 'The Neurologist as Ethics Consultant and as a Member of the Institutional Ethics Committee', *Neurological Clinics* 7, 697–713.

Fletcher, J. C. and Hoffmann, D. E.: 1994, 'Ethics Committees: Time to Experiment with Standards', *Annals of Internal Medicine* 120, 335–338.

Friedman, R. B.: 1973, 'On the Concept of Authority in Political Philosophy', in R. E. Flathman (ed.), *Concepts in Social and Political Philosophy*, Macmillan Publishing Co. Inc., New York, pp. 121–146.

Friedrich, C. F.: 1954, 'Loyalty and Authority', *Confluence* 3, 311–316.

Grunfeld, G. B.: 1990 , 'Non-MD Ethics Consultants?', *Journal of Clinical Ethics* 1, 325–326.

Hart, H. L. A.: 1961, *The Concept of Law*, Oxford University Press, Oxford.

Joint Commission for Accreditation of Healthcare Organizations: 1996, *Comprehensive Accreditation Manual for Hospital*, Joint Commission for Accreditation of Healthcare Organizations, Oakbrook Terrace, Illinois.

Jonsen, A. R.: 1992, 'Casuistry as Methodology in Clinical Ethics', *Theoretical Medicine* 13, 295–307.

LaPuma, J. and Priest, E. R.: 1992, 'Medical Staff Privileges for Ethics Consultants: An Institutional Model', *Quality Review Bulletin* 18, 17–20.

LaPuma, J. and Schiedermayer, D. L.: 1990, 'Must the Ethics Consultant See the Patient', *Journal of Clinical Ethics* 1, 56–59.

LaPuma, J. and Schiedermayer, D. L.: 1992, 'The Clinical Ethicist at the Bedside', *Theoretical Medicine* 12, 285–292.

Marsh, F. H.: 1992, 'Why Physicians Should not do Ethics Consultations', *Theoretical Medicine* 13, 285–292.

Morreim, H.: 1983, 'The Philosopher in the Clinical Setting', *Pharos* 46, 2–6.

Schutz, A.: 1967, *The Phenomenology of the Social World*, G. Walsh and F. Lehnert (trans.), Northwestern University Press, Evanston, Illinois.

Schutz, A., and Luckman, T.: 1973, *The Structures of the Life-World*, R. M. Zaner and H. T. Engelhardt Jr. (trans.), Northwestern University Press, Evanston, Illinois.

Thomasma, D. C.: 1991, 'Why Philosophers Should Offer Ethics Consultations', *Theoretical Medicine* 12, 129–140.

Veatch, R. M.: 1987, 'The Medical Ethicist as Agent for the Patient', in T. F. Ackerman, *et al.* (eds.), *Clinical Medical Ethics: Exploration and Assessment*, Lanham, NY, University Press of America, pp. 59–67

Veatch, R. M.: 1989a, 'Clinical Ethics, Applied Ethics, and Theory', in B. Hoffmaster, B. Freedman and G. Fraser (eds.), *Clinical Ethics: Theory and Practice* 1, Humana Press, Clifton, NJ, pp. 7-25.

Veatch, R. M.: 1989b, 'Advice and Consent', *Hastings Center Report* 19(1), 20–22.

Walker, M. U.: 1993, 'Keeping Moral Space Open: New Images of Ethics Consulting', *Hastings Center Report* 23(2), 33–40.

Weber, M.: 1947, *The Theory of Social and Economic Organization*, T. Parsons (tran.), New York, Free Press, New York.

Zaner, R. M.: 1984, 'Is "Ethicist" Anything to Call a Philosopher?', *Human Studies* 7, 71–90.

LORETTA M. KOPELMAN

CHANGING VIEWS OF PATERNALISM IN RESEARCH: AIDS ACTIVISTS DEMAND CHANGE

I. INTRODUCTION

It is a pleasure to write on the occasion of a *festschrift* to honor the retirement of my longtime colleague and friend Stuart Spicker. The topic that I have chosen reflects his well-known interest in the activities of institutional ethics committees and institutional review boards. Developed over the last part of the twentieth century, these committees have several well-defined goals. They seek to educate people within their organizations on relevant moral and social issues, assess the moral justifiability of their institutions' policies, and review troubling cases. Some of these committees evaluate the research conducted by their institutions and protect the rights and welfare of potential subjects in medical investigation, using the laws and social policies that have developed with the rising tide of medical research. There have been many changes in the area of medical research during Professor Spicker's career. I will discuss how these research policies, although developed with many good intentions, have been challenged on the grounds that they are too paternalistic. While many identifiable groups have questioned these regulations, the most vocal has probably been the acquired immunodeficiency syndrome (AIDS) activists, and so their concerns are my main focus. AIDS has caused many changes over the last part of this century and altering research policy is one of the most important. The fight against unjustified paternalism and concern for people with AIDS are also longtime concerns of Professor Stuart Spicker, making this an even more appropriate topic.

Systematic evaluation of medical therapies was rare until the middle of the twentieth century when physicians demanded scientifically grounded means for determining what helped their patients. Consequently, more and more patients were enrolled in controlled testing of therapies. Some of these patients were helped and some harmed, but unquestionably the scientific evaluation of therapies resulted in great medical advances for society. Another result of this increased research was the development of social policies to protect the rights and welfare

H. T. Engelhardt Jr. (ed.), The Philosophy of Medicine, 139–156.

of research subjects. These policies embody assumptions about how to gain informed consent and to protect people from harm. They also contain presuppositions about how to strike a balance between the importance of honoring people's rights, the social utility of conducting good research, and the need to protect people who may be enrolled in medical investigations. In devising these policies defenders and critics often disagree about how to strike this balance. When do patients have enough protection from the enthusiasms of investigators? Do patients have so many "rights" that investigators cannot conduct safe and important research? Discussion of these policies and presumptions about how to balance important values have grown until a chorus of defenses and criticisms of these policies now raise fundamental questions about how to plan studies and to select, inform, and treat subjects. This confrontation reveals the important role of values in every aspect of medical research, including its methods and goals, and shows that values are not peripheral to science but integral to its methods and conclusions.

In what follows, I will consider how AIDS activists and investigators disagreed about how to balance the social utility of research with the need to find justifiable means to protect the rights and welfare of patients or research subjects. More specifically, they disputed whether to alter research policy about the selection of treatments to test, the design of studies, the selection of subjects, when to begin or end trials, and who should have a role in making these decisions. The debate, in part, was and still is about expertise but it is also about power and control.

Many investigators and policy-makers insist patient and research subjects need many protections for their own good, even where they may not understand that they need it. Activists, however, demand more control in the planning and testing of promising new treatments; they decry the paternalism of investigators who say they know best about when to test, who to select, and when to release promising new treatments. The AIDS activists have been among the most vocal in making demands that the establishment be more responsive to their concerns, and reject investigators' claims that they best protect and assess subjects' interests (Kopelman, 1994a, 1994b, 1994c).

I will argue that resolution of this debate centers upon whether research can be a cooperative venture between investigators and subjects or the advocates they select. Research should be viewed as a cooperative venture because subjects and their advocates bring other interests and considerations than those investigators may consider (Kopelman, 1994a, 1994b,

1994c, portions of which are adapted here). In addition, supporting patients' civil liberties such as the right to consent has great social utility. Consequently, there are sometimes good reasons to reject past policies as too paternalistic. As we will see, however, some argue it cannot be a cooperative venture because subjects or their advocates make no meaningful contribution to decisions about when to begin, end, or release therapies to be public or because they tend to rely upon anecdotal information and so-called common sense rather than rigorous science (Annas, 1991; Kadane, 1986; Lacher, 1978; Tobias, 1988; Waldenstrom, 1983; Zelen, 1979). After clarifying the consensus about informed consent, the potential value conflicts and the meaning of "paternalism," I turn to how the AIDS activists are challenging the paternalism of policies on medical research.

II. THE CONSENSUS ON INFORMED CONSENT

At one time medicine was so paternalistic that doctors were expected to select people's therapies and unilaterally enroll them in studies. Since the late 1960s, a moral, legal, and medical consensus developed that competent adults' free and informed choices must be followed in devising treatment plans or enrolling them as research subjects (Brock, 1987; Faden, and Beauchamp, 1986). Ideally, patients (whether or not they are also subjects) share decision making with physicians as part of an ongoing process, where doctors clarify options and make recommendations. With few exceptions, competent adults' actual choices among these options must be followed even if doctors and nurses find them imprudent.[1]

According to the consensus in the moral and legal literature, informed consent for therapy requires appropriate disclosure of at least the nature of the patient's condition, the nature of the treatment recommendation, the alternatives, the consequences of these recommendations and alternatives, the consequences of no treatment, and the risks and likely benefits. The person's consent must be voluntary, or not coerced or manipulated. It must also be competent and this task-related notion means that the person can first, comprehend information relevant to the decision, second, deliberate based upon his or her values and goals, and third, communicate that information. The amount of information needed by the patient is generally understood as task-related. The person usually must be provided with sufficient information to understand and authorize or refuse a recommended plan. There is a duty to disclose all material information a physician knows, or should know, would be regarded as significant by a reasonable person in the patient's position when deciding to accept or reject a recommended medical procedure. The notion of informed consent is often summarized in terms of elements. Philosopher Tom Beauchamp and others have offered the following analysis: "(1) Disclosure, (2) Comprehension, (3) Voluntariness, (4) Competence, (5) Consent. Thus, one gives an informed consent to an intervention if and only if one receives a thorough disclosure about it, one comprehends the disclosure, one acts voluntarily, one is competent to act, and one consents to the intervention" (Beauchamp, 1991, 159S).

Consent for human *research*, given comparable risks, is generally stricter than consent for therapy because the goals are different. The goal of therapy is to help the patient, while the goal of human research is to gain knowledge. These goals are combined in therapeutic research, but the consensus holds that the primary goal should be the therapeutic benefit of the patient/subject, and the secondary goal to gain knowledge. Current regulations (U.S. 45 C.F.R. 46) for consent in research generally require that research subjects know about the nature, goals, risks, and possible benefits of studies. They must know if there are different groups and how they are chosen, as well as the institution's policy about compensating them if they are harmed. They must know whom to contact if they have questions about the study, and that they may withdraw at any time.

In short, a consensus has emerged about informed consent that subjects or patients must be at the center of decision making, if they are competent adults. The reasons include, first, that this is, in the long run, most socially useful because it strengthens our commitments to civil liberties generally to seek informed consent from people or their representatives for therapy and research. Second, it also seems to be in the best interest of most of us, since people are most likely to pay attention to and assess what benefits or harms them. Finally, self-determination is an extremely important moral and social value, and one that should be fostered by including, rather than excluding, people in decisions affecting them.

I will assume that violating the consensus on informed consent is wrong. This posture does not mean that failing to gain informed consent is actually wrong, but that there is a burden of proof upon those who waive or modify consent standards to justify their action. Overcoming this presumptive duty to gain informed consent in therapy may be accomplished by demonstrating that the study or treatment constitutes a well-established exception to the duty to gain consent. Faden and Beauchamp enumerate such exceptions as, "the public health emergency, the medical emergency, the incompetent patient, the therapeutic privilege, and the patient waiver. All but one of these exceptions, the controversial therapeutic privilege, are commonly taken as both morally and legally valid, and thus as being consistent with – or placing valid limits on – the duty to respect individual autonomy (Faden and Beauchamp, 1986, p. 35). With the exception of the rare public health emergency, these established exceptions generally do not apply to research.

Overcoming this presumption to gain informed consent in research generally requires showing that the risk to subjects is so minimal few subjects would object, and that the difficulty of gaining consent is sufficient to discourage important but safe research. But no laws can be violated, and subjects must be debriefed whenever possible. The regulations also occasionally allow investigators to omit consent requirements for "research reviewing the collection or study of existing data, documents, records, pathological specimens, or diagnostic specimens, if these sources are publicly available or if the information is recorded by the investigator in such a manner that subjects cannot be identified directly or through identifiers linked to the subject" (U.S., 1991, 45 CFR 46.101(b)(4)).

For purposes of this discussion, I would like to classify two distinct categories of assessments of choice regarding people's informed consent. Let us call what people are willing to authorize *descriptive* assessments of their choices. In contrast, let us classify what experts think they ought to authorize *normative* assessments. Using these categories, the recent consensus on informed consent recasts treatment and research authorization from a largely normative notion (what experts believe people ought to want) to a generally descriptive concept (what they actually do want). This evolution is roughly analogous to general movement from paternalism toward patient autonomy.

III. VALUES

There are many important and potentially conflicting values that need to be considered in justifying research, and over the years a certain balance has been struck in assessing the justifiability of conducting research with human subjects. First, medical research has great social utility. Research improves diagnosis, therapy, and prevention of disease for everyone. Without good research, moreover, there is little information about the benefits and risks of alternative treatments. In short, unless there are good studies, physicians have difficulty vindicating their therapeutic intentions or obtaining good information.

Social utility, however, cannot be the final appeal about what studies should be done. We also need to gain consent, since gaining consent from people or their representatives serves a second important value: the just treatment of persons. Consent honors them as persons capable of rational self-determination. It shows we have respect for them not to be used as objects of study without their consent. Seeking consent, however, can inhibit medical progress. Consequently, we sometimes need to rank the relative importance of gaining informed consent and promoting social utility through studies. Respect for persons' rights generates the duty to gain consent and avoid paternalism. This needs to be balanced against the social utility of research in securing generalizable knowledge and medical advances.

There is, however, a third value that needs to be a part of the balance. In addition to the social utility of studies and the need to respect people's rights, a third value recognizes that people may need protection. Gaining consent offers one important way to protect people. People are usually most interested in protecting their own well-being and that of the people for whom they are responsible. They do not want to lose opportunities or suffer unacceptable procedures or risks. However, many people lack information, so the good will of investigators is also an important means of protection of human subjects. Because a small number of abuses revealed that this was not sufficient (Beecher, 1966), other important means of protection of subjects were devised including the requirement to have most human research protocols approved by institutional ethics committees or institutional review boards using national guidelines.

A fourth important value shaping research policies concerns the fair selection of subjects. To draw subjects predominately from certain economic, racial, or special groups would be unfair. Since research benefits everyone, it is wrong to place the risks on one group. For example, this policy prohibits having poor people alone serve in dangerous studies. On the other hand, it can also be unfair to exclude certain groups as subjects who might otherwise be eligible and benefit.

For example, it is wrong to exclude women from serving as subjects without justification since participation maybe their best chance for good care, and benefits women collectively (Dresser, 1992, Wenger *et al.*, 1993).

Gaining consent and opposing unjust paternalism helps promote trust, cooperation, and understanding. It works toward thinking of research as a cooperative venture where all these interests are balanced by a reasoned and open consensus. This does not mean every agreement is correct, of course, but that a public debate is more likely to represent everyone's interests adequately. In our wider social policies, unless there is an extraordinary event such as a public health emergency, we generally regard values of informed consent, protection from unjust paternalism, and fairness in selecting subjects as more important values than medical advances.

IV. PATERNALISM

Paternalism suggests that a person takes an interest in someone as a devoted father caring for his child. Thus paternalism, or "parentism," as some prefer to call it, means acting for someone's best interest, and implies that the person *needs* protection as a child needs the protection of a caring parent. Adults who are incompetent because they are unconscious or mentally retarded need the protection of others who should make decisions about what is best for them. For example, a police officer acts paternalistically when preventing someone from committing suicide, and a physician does so in protecting an abused child. Police and physicians have special duties in these cases, but even strangers have duties to help others and report such things as child abuse.

When "paternalism" is used in this *wide* way, it can be justified or unjustified. Justifiable paternalism presupposes 1) the act is intended to be beneficent and 2) the person has special insight, experience, skill, authority, responsibility, or knowledge to justify intervening as a loving father taking care of his child (Beauchamp and McCullough, 1984). It is unjustified paternalism if the intention is not beneficent or if the person lacks insight, knowledge, authority, responsibility, experience, or skill to justify the intervention. I will use "paternalism" in this wide way because it follows common usage and is reflected in the law.

Some prefer to use "paternalism" in a *narrower* way, however, so that paternalism is always unjustifiable. Even though I will not use it in this way, I point this out to avoid the confusion that often follows from confounding them. The narrower sense limits the meaning of "paternalism" to its pejorative meaning so that all acts of paternalism are wrong. That is, paternalism is simply defined as unfairly overruling competent people who can or should make choices for themselves about what is best for them (Beauchamp and McCullough, 1984). I will not use "paternalism" in this narrower way for several reasons including its difference from common usage.[2]

Some critics argue that investigators and policy makers are unjustifiably paternalistic in some of their decisions about who can participate in research or in establishing certain rules about the testing and distribution of promising new medical interventions. In what follows, I will discuss examples of such claims and why investigators risk losing public trust if they do not respond to charges of unjustified paternalism.

V. AIDS ACTIVISTS

Over the last generation, the views of activists and investigators increasingly collided over how to control the testing and use of promising new treatments. This is nowhere better illustrated than in the example of activists seeking the best possible treatments for people with human immunodeficiency virus (HIV) infection. These activists argued that despite investigators' and policy makers' good intentions, people's interests are sometimes needlessly sacrificed to rigid, narrow, inappropriate, paternalistic or over-demanding research procedures (Dresser, 1992; Kopelman, 1994a, 1994b; Clayton *et al.*, 1995; Fried, 1974; Lo, 1992; Wenger, 1993).

Activists object to the degree of control that investigators sometimes have over the lives of people who are very sick and have few options. Consider their criticisms in a wider context of social liberties. People in this country may do any number of foolish and life-threatening things without permission from the government. Adults are at liberty to drive race-cars, hang-glide, bungee-jump, and climb into active volcanoes, so why should they be restricted "for their own good" when it comes to gaining access to treatments that may genuinely help them? That is, why should they need its protection to pay for or take a drug that they believe (sometime with good reason) will help them?

Testing by the Food and Drug Administration (FDA) is not only excessively slow, critics charge, but the FDA too tightly controls who can receive untested or experimental therapies. Several issues were specifically cited as unfair: First, they held it to be unreasonable that only those enrolled in clinical trials (CTs) of investigational new drugs (INDs) can usually receive them. Second, they challenged the fairness of automatically excluding certain groups from participation in research trials such as women, children, and those with serious complications. It is unfair especially when CTs are free or offer the only hope for good treatments. Third, exclusionary rules do not make sense for rapidly progressing fatal illnesses. Fourth, they argue that competent people should have more choices about whether they are willing to undergo risky, experimental, or even untested therapies.

A. *Defenders of Traditional Clinical Trials Methods*

One group, composed largely of investigators, have generally resisted the changes sought by activists to make experimental and relatively untested interventions more generally available to the public; they argue that research cannot be a cooperative venture between investigators and subjects or their representatives (Annas, 1991; Kadane, 1986; Lacher, 1978; Tobias, 1988; Waldenstrom, 1983; Zelen, 1979). These modifications undercut proven research methods; changing the procedures for testing and availability of new treatments because of activists' pressure, might help a few, but will harm many in the end if untested or poorly tested therapies become widely available. They doubt the meaningfulness of informed consent and whether potential subjects can understand enough about risks or benefits of alternative treatments, randomization, interim data, probabilities, or the scientifically rigorous methodology used in testing to make them anything like partners. Some asserted that the problem is that patients, families, and their advocates tend to rely upon anecdotal information and so-called common sense (Annas, 1991). Research cannot be a cooperative venture between investigators and subjects or their advocates, they argue, because they cannot understand enough to be a party to the key decisions about when to begin, end, or release promising new therapies or devices. Rather, we must defend investigators' freedom to design what they regarded as the best studies and protections. People cannot defend themselves adequately through informed consent, they argue, so the most knowledgeable clinician-investigators should protect the public from dangerous and ineffective "therapies." The investigators argue that the best way to fulfill duties, to improve treatments, and protect the public is to give investigators and their peers the freedom to set the goals, priorities, and methods for testing promising new treatments. Patients have a right to optimal treatment, they maintain, but not to pick their treatment (Zelen, 1979). The current understanding of patients' rights is unreasonable, they hold, and disrupts clinical trials. If investigators give up still more control, they claim we would further slow medical progress and muddy the waters with even worse methodology. They conclude that we should support investigators' freedom to design the best clinical trials and modify patients' rights for the sake of public utility.

Activists pressing for greater rights and control are not likely to be satisfied with fewer rights, even given these assurances that future patients will benefit from better clinical trials. These activists do not agree that investigators make the most socially useful choices. The "right" to design trials without interference, moreover, is shallow if patients will not enroll in them. Investigators should not underestimate people's interests and ability to learn what is going on, especially if the research involves risks or a disease they follow closely. Potential subjects may feel tricked by a lack of candor if they find out their consent was manipulated or omitted. While many potential subjects cannot understand much of what they are asked to do or why, it is a mistake to assume none can. This solution, therefore, is not likely to be popular outside the research community.

B. *Critics of Traditional Clinical Trials Methods*

Another group agrees with these investigators in questioning if research can be a cooperative venture where subjects or their advocates are seen as partners in clinical trials (Fried, 1974; Gifford, 1986; Kadane, 1986; Marquis, 1986; Schaffner, 1986; Wikler, 1981). They also agree that most people cannot understand enough to be partners and that the public must be protected. They prefer to restrict the kinds of studies that can be done, however, rather than give investigators and their peers wide discretion. That is, while agreeing that research can not be a cooperative venture with patient-subjects because patients do not understand enough to be partners, they reached a different conclusion about what policy to adopt.

Investigators do not tell, and most patients do not understand, they maintain, that at some point in the trial, some groups generally are known to get sub-optimal care (Wikler, 1981). Clinical trials, they argue, put medical advances ahead of individual patients' rights and welfare. The patient should come first, and doctors should not enroll their patients in studies where they do not come first. Traditional research methods, they hold, typically violate physicians' duties to their patients. These critics understand the physicians' duty, or what some call the "therapeutic obligation," as the duty to provide patients with what their doctors believe is the best available care. Traditional clinical trial methodology, they hold, entails that some patients receive sub-optimal care, and conclude such methods are incompatible with the traditional duties of health care professionals.

This statement of physicians' duty or the therapeutic obligation, however, presumes that there is a best treatment and it is up to physicians to determine what is the best treatment for their patients. Both assumptions are often false. First, one reason for engaging in clinical trials is uncertainty about which treatments are best. Thus, none can assert what the best treatment is in such circumstances. That is, a moral requirement for justifying clinical trials is that the different arms of the study must be in clinical equipoise. This is usually understood to mean that the community of investigators and physicians are uncertain about which treatment being tested is best, so there is no known therapeutic advantage to a patient's assignment to any one of the various treatment arms (Freedman, 1987). Claiming physicians must provide what they believe is the best treatment available for their patients, therefore, assumes that they know what is the best treatment. Yet, this is incompatible with the assumption underlying clinical trials that the best treatment is unknown. It is not surprising, therefore, that these critics deny the morality of this sort of testing since they assume treatments cannot be in clinical equipoise.

Second, this formulation of the physicians' duty or therapeutic obligation (that physicians must provide patients with what doctors believe is the best available care) is paternalistic because it assumes that physicians know what is best for patients. Yet, AIDS activists have demonstrated they are unwilling to let others decide what is best for them. Many patients want to consider treatments in light of their own goals, values, and principles to determine how alternative treatments affect the quality of their lives. If patients do not think the treatment arms of the trial are in equipoise from their perspective, they will refuse to participate. Lo argues that in addition to investigators conducting clinical trials, potential volunteers should also help decide what treatments are in equipoise (Lo, 1992).

To summarize, critics argue that research cannot be a cooperative venture because patients cannot understand enough to be genuine partners. They conclude that as a result, we have to make some hard choices about how to protect the public. Some of these critics favor giving investigators' freedom to design what they consider to be the best trials and others seek to protect patients' rights or welfare by greater restriction, regulation, or oversight regarding what studies can be done.

C. *Rejection of Protectionism*

We have considered two groups who hold that research cannot be a cooperative venture. They disagree about whether to defend traditional clinical trial methodologies. We found both views very problematic. Moreover, activists want a greater role in setting priorities and in saying how studies are conducted, so not surprisingly they reject both proposals for increased protectionism. They draw attention to the evaluative and ethical nature of many questions about the testing and use of promising new treatments. These evaluative and ethical issues concern decisions about the initiation of trials, the use of placebos, randomization, the selection of subjects, the ending of trials, and how best to protect the general public with respect to promising new treatments. These are not matters that can be settled entirely by science, since they contain many critical and controversial value choices. Activists want a say in resolving them, insisting that however good the techniques devised by investigators to protect the public from untested and dangerous drugs may be, they can be improved.

VI. CHANGES

Investigators gradually began to make changes as a result of activists demands or because investigators saw that their efforts could be defeated by subjects who do not agree with the goals of studies. If subjects are not convinced of studies' merits they will not enroll; or, if they do enroll, they may refuse to follow investigators' directions. For example, Merigan found that subjects can defeat trials by not following the investigators' instructions. He reported that the clinical trial of zidovudine (AZT) against a placebo was jeopardized by the fact that 9% of the patients who were given a placebo were taking AZT on their own. As Merigan writes, "If these irregularities had been more widespread, the useful effects of the drug could have been obscured in the trial, to the detriment of future patients infected with HIV. Specifically, the early initiation of zidovudine therapy in asymptomatic patients would not have become standard practice" (Merigan, 1990, 1342).

Patients can also defeat trials by refusing to enroll in them. For example, the National Surgical Adjuvant Project for breast and bowel cancers set out to conduct randomized clinical trials on the survival rates of women with breast cancer. Women were asked to give consent before randomization to lumpectomy versus simple mastectomy with or without radiation. Women were asked to go into surgery not knowing if they had cancer, and if they had it, not knowing which treatments they would receive. Because of this uncertainty many physicians were reluctant to ask their patients to participate in these studies; doctors who agreed to ask them found many women unwilling to accept such conditions (Ellenberg, 1984; Fisher, 1980; Taylor, Margolese and Soskolne, 1984). In response to the slow accrual rate, investigators switched to a prerandomized schema. Subjects were informed of their proposed group assignment, treatment options, and the nature and purpose of the study at the time consent was sought. Accrual rates increased sixfold. Several authors found this suspicious and questioned if physicians, perhaps unconsciously, were enthusiastic about whatever assignment the women got since they knew the assigned treatment in advance of seeking consent (Beauchamp and Childress, 1989; Ellenberg, 1984; Marquis, 1986).

These critics of prerandomization may discount another important explanation. Patients may find prerandomization design more acceptable because they know what treatment they will receive. It is not that they cannot understand randomization. Randomization is no harder to understand than the flip of a coin. The problem with randomization is related, I believe, to how we react to uncertainty when we are very apprehensive. When there is great potential loss or gain, agreeing to randomization forces us to confront uncertainty. We cannot rationalize that we have control or that there is some rational way to break the tie. The issue is not merely about information, but our difficulty in admitting that we have little control. Randomization forces us to admit we have no defenses against our uncertainty, risk, and loss of control.

In response to these concerns and criticisms of protectionism changes occurred. For example, surrogate or "soft" endpoints sometimes replaced hard endpoints in ending trials. For example, they were used in approving dideoxyinosine (DDI) for marketing. Investigators did not know how DDI affected the "hard" endpoints, death, or opportunistic infections; they did know it increased lymphocyte levels. The use of soft or surrogate endpoints to end trials satisfied those who wanted to speed up trials, but brought criticism from those who questioned whether we understood enough about the relationship between the hard and soft endpoints.

In addition, in an attempt to defuse some of these problems, the Public Health Service (PHS) expanded the availability of investigational new drugs by means of a parallel track. Patients with HIV-related diseases can obtain experimental drugs even though they are not enrolled as subjects in the clinical trials of the drug. These regulations require that there are no therapeutic alternatives, the drugs are being tested, there is some evidence of their efficacy, there are no unreasonable risks for the patient, and the patient cannot participate in the clinical trials (U.S., PHS, 1990).

Consequently, many more doctors could include their patients in trials if investigators streamlined data sets or asked fewer, more pertinent questions. In addition, investigators could speed up testing by using factorial designs that compare two or more drugs simultaneously. The use of more flexible entry criteria and streamlining of data sets would encourage more collaboration with physicians around the country. This would have the consequence of enabling more patients to enroll, making results more generalizable. It would also make treatments available to more patients and minimize the use of parallel treatment tracks.

In May of 1992, a promising new antiviral agent, D4T (Stavudine), began Phase II/III testing. This testing illustrates these changes. First, Phase II testing was not completed before Phase III began. Second, it is being compared with AZT and physicians can obtain D4T for their HIV patients on a parallel track free from the manufacturer. It did, however, exclude two groups of children under 13 years and women who are pregnant or breast-feeding infants. This same manufacturer distributed DDI (trade name didanosine) to over 23,000 patients on expanded-access programs.

Not everyone is happy about these changes. Some do not think they go far enough. For example, as we saw in the testing of the agent D4T (Stavudine), pregnant women and children under 13 were excluded from phase II/III of testing. Some in these groups might continue to decry restrictions. On the other hand, some object that these new policies go too far and believe they risk public safety by undercutting proven methods and scientific standards. The general public may not understand that drugs available on a parallel track are largely untested, or why insurance companies refuse to pay for untested therapies. Moreover, members of the public may not participate in clinical trials once they realize the promising new drug is available outside the study. Investigators and policy makers are also under attack for the way in which they select and protect certain groups of subjects.

VII. CONCLUSIONS

These examples raise troubling questions about unjustified paternalism in medical research. Paternalism is unjustified when measures designed to protect people do not, when competent and informed people do not want the "protection," or when the motive is not really beneficent. I have considered and rejected several different reasons for holding research cannot be a cooperative venture between subjects or their representatives and investigators. Some good charges have occurred. Investigators must now give cogent explanations for excluding groups such as women, children, or racial minorities; the public now has access to elaborate web sites at N.I.H. where people can find reliabe and timely information. Unjustified paternalism erodes public trust especially when the public lacks conviction that all interests are adequately represented. If these issues are not addressed adequately, the public may lose confidence in choices made by investigators. Investigators frequently argue that the long-term benefits of careful research justify keeping current policies. They need to consider, however, why the public is increasingly suspicious of the enterprise of medical research. The authority that some investigators cling to, moreover, could easily be lost in the social and political arena. The damage to the public's trust in research as a cooperative venture where everyone's interests are adequately represented could result in fewer willing subjects, less funding, more oversight, and greater restrictions. Unless the research community assures the public that research is a cooperative venture, it risks losing something more important than comfortable or favored models: it risks losing public confidence.

Department of Medical Humanities
East Carolina School of Medicine
Greenville, North Carolina, U.S.A.

NOTES

[1] Where the subject or patient is incompetent the person's representative, such as a parent, gives consent. Their views need not be controlling if the professionals believe their choices are abusive or neglectful.
[2] This usage also is tied to specific moral theories and assumptions about what autonomy means.

BIBLIOGRAPHY

Angell, M.: 1993, 'Caring for Women's Health: What is the Problem?', *New England Journal of Medicine* 329(4), 271–272.

Annas, G. J.: 1991, 'AIDS, Compassion and Drugs', *Hastings Center Report* 21(4), 44–45.

Ayanian, J. Z., Udvarhelyi, G., Pashos, C. L., and Epstein, A. M.: 1993, 'Racial Differences in the Use of Revascularization Procedures after Coronary Angiography', *Journal of the American Medical Association* 269 (May 26), 2642–2646.

Ayanian, J. Z. and Epstein, A. M.: 1991, 'Differences in the Use of Procedures between Women and Men Hospitalized for Coronary Heart Disease', *New England Journal of Medicine* 325, 221–225.

Beauchamp, T. L., *et al.*: 1991, 'Ethical Guidelines for Epidemiologists', *Journal of Clinical Epidemiology* 44S(1), 151S–169S.

Beauchamp, T. L. and McCullough, L.: 1984, *Medical Ethics*, Prentice Hall, New York, Chapter 4.

Beauchamp, T. L. and Childress, J. L.: 1989, *Principles of Biomedical Ethics*, 3rd edition, Oxford University Press, Oxford.

Beecher, H. K.: 1966, 'Ethics and Clinical Research', *New England Journal of Medicine* 274(24), 1354–1360.

Brock, D.: 1987, 'Informed Consent', in D. Van De Veer and T. Regan (eds.), *Health Care Ethics*, Temple University Press, Philadelphia, pp. 98–124.

Clayton, E. W., Steinberg, K. K., Khoury, M. J., Thomason, E., Andrews, L., Kahn, M. J. E., Kopelman, L. M., and Weiss, J. O.: 1995, 'Informed Consent for Genetic Research on Stored Tissue Samples', *Journal of the American Medical Association* 274(22), 1786-1792.

Dresser, R.: 1992, 'Wanted: Single, White Male for Medical Research', *Hastings Center Report* 22(1), 24–29.

Ellenberg, S. S.: 1984, 'Randomization Designs in Comparative Clinical Trials', *New England Journal of Medicine* 310, 1404–1408.

Faden, R. R. and Beauchamp, T. L.: 1986, *History and Theory of Informed Consent*, Oxford University Press, New York.

Fisher, B.: 1980, 'The National Surgical Adjuvant Project for Breast and Bowel Cancer', *NSABP Protocol B-06*. Distributed in 1980 to surgeons.

Freedman, B.: 1987, 'Equipoise and the Ethics of Clinical Research', *New England Journal of Medicine* 317, 141–145.

Fried, C.: 1974, 'Medical Experimentation: Personal Integrity and Social Policy', in A. B. Bearn, D. A. K. Black and H. H. Hiatt (eds.), *Clinical Studies* 5, Elsevier Press, New York.

Gifford, F.: 1986, 'The Conflict between Randomized Clinical Trials and the Therapeutic Obligation', *Journal of Medicine and Philosophy* 11, 347–366.

Kadane, J. B.: 1986, 'Progress Toward a More Ethical Method for Clinical Trials', *Journal of Medicine and Philosophy* 11(4), 385–404.

Kopelman, L. M.: 1994a, 'How AIDS Activists are Changing Research', in J. R. Monagle and D. C. Tomasma (eds.), *Health Care Ethics: Critical Issues*, Aspen Publishing Co., Gaithersburg, Maryland, pp. 199–209.

Kopelman, L. M.: 1994b, 'Research Methodology/II. Controlled Clinical Trials', in W. T. Reich (editor in chief), *Encyclopedia of Bioethics*, vol. 4, 2nd edition, Macmillan Free Press, New York, pp. 2270–2285.

Kopelman, L. M.: 1994c, 'Informed Consent and Anonymous Tissue Samples: The Case of HIV Seroprevalence Studies', *Journal of Medicine and Philosophy* 19, 525–552.

Lacher, M. J.: 1978, 'Physicians and Patients as Obstacles to a Randomized Trial', *Clinical Research* 26, 375–379.

Lo, B., 1992, 'Ethical Dilemmas in HIV Infection: What Have We Learned', *Law, Medicine and Health Care* 20, 92–103.

Marquis, D.: 1986, 'An Argument that all Prerandomized Clinical Trials are Unethical', *Journal of Medicine and Philosophy* 11(4), 367–384.

Merigan, T. C.: 1990, 'You Can Teach an Old Dog New Tricks: How AIDS Trials are Pioneering New Strategies', *New England Journal of Medicine* 323(19), 1341–1343.

Merkatz, R. B., Temple, R., Sobel, S., Feiden, K., and Kessler, D.: 1993, 'And the Working Group on Women in Clinical Trials: Inclusion of Women in Clinical Trials – Policies for Population Subgroups', *New England Journal of Medicine* 329(4) (July 22), 288–296.

Schaffner, K.: 1986, 'Ethical Problems in Clinical Trials', *Journal of Medicine and Philosophy* 11(4), 297–315.

Steingart, R. M., Packer, M., Hamm, P., *et al.*: 1991, 'Sex Differences in the Management of Coronary Artery Disease', *New England Journal of Medicine* 325, 226–230.

Taylor, K. M., Margolese, R. G., and Soskolne, C. L.: 1984, 'Physician's Reasons for not Entering Eligible Patients in a Randomized Clinical Trial of Surgery for Breast Cancer', *New England Journal of Medicine* 310, 1363–1367.

Tobias, J. S.: 1988, 'Informed Consent and Controlled Trials', *Lancet* 2:8621, 1194.

Todd, K. H., Samaroo, N., and Hoffman, J. R.: 1993, 'Ethnicity as a Risk Factor for Inadequate Emergency Department Analgesia', *Journal of the American Medical Association* 269 (March 24), 1537–1539.

Todd, K. H., Lee, T., and Hoffman, J. R.: 1994, 'The Effect of Ethnicity on Physician Estimates of Pain Severity in Patients with Isolated Extremity Trauma', *Journal of the American Medical Association* 271(12), 925–928.

United States, 'Protection of Human Subjects', 45 C.F.R. sec. 46 (1991).

U.S. Department of Health and Human Services: 1980 (May 30), 'Protection of Human Subjects; Prisoners Used as Subjects in Research', *Federal Register* 45, 36386–36391; 46, 306.

U.S. Department of Health and Human Services: 1981 (Jan. 26), 'Final Regulations Amending Basic HHS Policy for the Protection of Human Subjects', *Federal Register* 46, 8366–8392.

U.S. Department of Health, Education and Welfare: 1978 (Nov. 16), 'Additional Protection Pertaining to Biomedical and Behavioral Research Involving Prisoners as Subjects',

Federal Register 43, Part IV:HEW 53652–56 and DHEW, Food and Drug Administration.

U.S. Public Health Service Consultation on International Collaborative Human Immunodeficiency Virus (HIV) Research. Reprinted in Law, *Medicine and Health Care* 19(3–4), 259–263.

Udvarhelyi, I. S., Gatsonis, C., Epstein, A. M., Pashos, C. L., Newhouse, J. P., and McNeil, B. J.: 1992, 'Acute Myocardial Infaction in the Medicare Propulation: Process of Care and Clinical Outcomes', *Journal of the American Medical Association* 268 (Nov. 11), 2530–2536.

Wikler, D.: 1981 (Dec. 8), 'Ethical Considerations in Randomized Clinical Trials', *Seminars in Oncology*, 437–441.

Waldenstrom, J.: 1983, 'The Ethics of Randomization', in *Research Ethics*, Alan R. Liss, Inc., New York, pp. 243–249.

Wenger, N. K., Speroff, L., and Packard, B.: 1993, 'Cardiovascular Health and Disease in Women', *New England Journal of Medicine* 329, 247–256.

Zelen, M.: 1979, 'A New Design for Randomized Clinical Trials', *New England Journal of Medicine* 300(22), 1242–1245.

SECTION THREE

FUNDAMENTAL CATEGORIES: THE MIND, EQUITY, THE STATE, AND TIME

K. W. M. FULFORD

THREE DESIGNATIONS OF DISORDER: DIVERSITY, DISEASE AND DETERMINISM IN PSYCHIATRIC THOUGHT AND PRACTICE

I. INTRODUCTION

Psychiatry is the despair of the academically tidy-minded. Ostensibly it is part of medicine; but it lacks a unified nosology, its research base is in sociology, anthropology and psychology as much as in the biological sciences, and its treatment methods range from drugs and electric shock through cognitive and behavioral techniques to counseling, psychotherapy and psychoanalysis. Worse still, the very concept of mental illness, contrary to the scientific self-image of medicine, is heavily value laden. Indeed, psychiatry is so problematic ethically, that many have taken mental illness to be a moral rather than a medical concept (Szasz, 1960 and 1987). Small wonder, then, that as a medical discipline psychiatry has been and remains under attack on a number of fronts (Clare, 1979).

What are we to make of this disorderly discipline? In this article, I examine psychiatry from three perspectives: historical, conceptual and bioethical. The common theme linking these perspectives is the medical model. Historically, the diversity of psychiatry's theoretical foundations reflects a series of only partially successful attempts to establish a medical model of mental illness comparable with that available for physical illness. Conceptually, much of the debate about the validity of the concept of mental illness turns on the extent to which the determinism implicit in the medical model is appropriate in psychiatry. Bioethically, the widespread assumption of an essentially medical model of mental illness has led to a relative neglect of psychiatry by bioethicists.

Recognizing all this repolarizes our view of psychiatry. Instead of being the runner-up to medical science, instead of being a conceptually dubious shadow side of physical medicine, instead of being a postscript to bioethics, it emerges at the cutting edge of a new philosophy of healthcare. This philosophy, although derived here from psychiatry, has a number of important features in common with modern work in the philosophy of medicine, notably that of the subject of this celebratory

H. T. Engelhardt. Jr. (ed.), The Philosophy of Medicine, 159–176.

volume, Stuart Spicker. It is a philosophy equally of mind *and* body, therefore, pointing forwards to a model of healthcare which is better tuned than the traditional medical model to the scientific as well as the human demands of medicine in the twenty-first century.

II. A DIVERSITY OF HISTORICAL THEMES

As a predominantly medical specialty modern psychiatry has its origins in the dramatic developments in medical science which took place in the nineteenth and early twentieth centuries. Mental illness was not invented then. In many countries, and as far back as classical times, insanity has been thought of as a disease (Kenny, 1969). There have been competing theories: depravity, possession, inspiration. But over this period a number of factors combined to reinforce the disease theory. The rationalism of the French school of psychiatry, dominant in the first half of the century, was associated with a firmly naturalistic view of madness. Most of its leading figures – Pinel and Esquirol for example – were physicians; and the humanitarian reforms in the treatment of the mentally ill introduced at this time were guided in part by a medical model of madness. This in turn ensured a prominent place for the medical profession in the management of the lunatic asylums which sprang up in Europe and America in association with the expanding cities of the industrial revolution. A new breed of medical experts thus emerged with specialized knowledge and experience of mental disorders. And their pre-eminence was secured by association with the successes of the germ theory and other contemporaneous advances in medical science, for these held out the prospect, perhaps for the first time, of cure, rather than merely containment, of the mentally ill.

From the start there was considerable diversity of opinion about mental disorder. Almost every expert had his own classification (Kendell, 1975a)! This was itself a product of the medical model, however, the different classifications reflecting competing etiological theories. In the event all these theories proved to be premature, a degree of order emerging only with the publication of Emil Kraepelin's classification in the sixth edition of his *Lehrbuch* in 1899. Kraepelin, too, believed in specific mental disease entities, "lesions" defined by changes in bodily structure and functioning. But he recognized that the first step towards discovering such entities was the careful delineation of syndromes defined by the symptoms and time course of mental disorders. Kraepelin thus took psychiatry, on the model of physical medicine, back to an earlier Sydenham-like stage of classification. His hopes for specific disease entities have still to be realized, though recent developments in brain imaging techniques have given a new stimulus in this direction. But the main outlines of Kraepelin's classification, together with the careful phenomenological studies of Karl Jaspers (Jaspers, 1913a), remain the basis of modern descriptive classifications of mental disorders. Moreover a broadly organic understanding of mental illness has in part been vindicated by the appearance in the 1950s and 1960s of powerful psychotropic drugs.

Yet even as Kraepelin's system, implying a biological disease model of insanity, was becoming established, two other approaches were making their appearance in European psychiatry: the psychoanalytic, with Freud in Vienna, and his theory of mental disorder as a product of largely unconscious mental processes distorted in development through adverse experiences and failures in early relationships; and the psychological, with Pavlov in St. Petersburg, and theories of the neuroses based on maladaptive conditioning and learned abnormalities of behavior. For several decades these approaches were to a large extent marginalized by mainstream psychiatry. Both were influential in certain countries: Freud in America, Pavlov in Russia, for instance. Similarly, both were influential outside medicine: Pavlov in psychology; Freud and his successors in the cultural and literary traditions of Europe and America (Storr, 1989). But in recent years psychology and psychoanalysis have come to play increasingly important roles in psychiatry itself. Psychological techniques, behavioral and cognitive, have become established as treatments of choice, superior to psychotropic drugs, for a wide range especially of neurotic disorders. Psychoanalysis, though still not widely available in the U.K. in its own right, has become highly influential in child psychiatry, in the development of counseling and psychotherapeutic services, in the therapeutic community movement, and as one of the roots of the relatively new discipline of social psychiatry (Holmes and Lindley, 1989).

Present day psychiatry thus embraces a number of to some extent competing theoretical models. The psychiatrist remains a medically qualified practitioner, but working within a multi-disciplinary team, and drawing on a wide variety of different skills and resources in his or her day-to-day clinical work. Yet the elements of modern psychiatry, diverse as they are, have an important feature in common. They are all deterministic. This is widely acknowledged, indeed it is seen as a strength, of the disease models of biological psychiatry (Christen and Churchland, 1992). But broadening the medical model to a more "biopsychosocial" paradigm, although in many ways helpful, fails to avoid the deterministic net. Pavlovian conditioning, and later psychological paradigms, are, literally, behaviorist. Psychoanalysis, though concerned with persons, and written in the language of intention and meaning, is predominantly deterministic in outlook. Freud's theories are of psychological causation. Even social psychiatry has been developed largely on a positivistic model of social science (Wing, 1978).

There have been anti-determinist movements in psychiatry, to be sure. Adlers' emphasis on the importance of the individual was to some extent a reaction to the reductionism of psychoanalytic theory. Jaspers too, and indeed Kraepelin, wrote of the need for meaningful as well as causal explanation in psychology (see, for example, Jaspers, 1913b). Existential analysts, and some of the neo-Freudians (Fromm and Sullivan, for example) have emphasized our capacities as individuals with freedom of thought and action. But all these are marginalized still. The dominant theoretical models of medical psychiatry share a clear central assumption of scientific determinism.

III. DISEASE MODELS AND THE CONCEPT OF MENTAL ILLNESS

The determinism implicit in psychiatry's theoretical models is explained partly by practical and partly by conceptual considerations. To the extent that determinism is equated with scientific thinking, it reflects the practical success of the medical model. Medicine is a pragmatic discipline. If deterministic models yield results then deterministic models are "in." The proven value of science in physical medicine is thus a sufficient justification for at least the attempt to extend a deterministic disease model to psychiatry. As noted a moment ago, this has already led to important advances, though psychiatry is still the least developed of the scientific medical specialties.

The conceptual justification for determinism in psychiatry, however, being essential rather than merely contingent, is more significant. What it amounts to is that the very concept of disease (whether construed biologically, psychologically or psychoanalytically) implies a loss of that freedom which, intuitively, is the antithesis of determinism. The very term patient implies a loss of agency. To be ill is to be incapacitated in one respect or another. It is also to be, to a greater or lesser extent, not responsible. This is true in law: epileptic automatism is an excuse, for example. It is also true in everyday contexts: being "off sick" justifies absence from work, in contrast to malingering. This feature of illness has been emphasized particularly by sociologists. Talcott Parsons, for example, included it as a key element in the distinction between disease and other forms of deviance (Parsons, 1951). To the extent, then, that psychiatry is concerned with a disease model of mental illness, there is a conceptual as well as practical justification for its broadly deterministic stance.

Yet it is this same determinism which is behind much of the criticism to which the concept of mental illness has been subject in recent years. Szasz was among the leaders of the anti-psychiatry movement. He argued, in effect, that mental illness is an illegitimate extension of a purely descriptive (and hence scientific) concept of physical illness (Szasz, 1960 and 1987). But his underlying concern was to return the locus of control to the patient. So-called mental illnesses, he claimed, are really "problems of living" for which we are fully responsible. Despair is despair, not disease. Delinquency is delinquency. Hysteria really is just malingering. Similarly, though from a different political orientation, Foucault argued that dissent is dissent. The medicalization of madness, he suggested, was a nineteenth-century device, born of the needs of the industrial revolution, designed to invalidate protest and to justify the incarceration of the mentally ill (Foucault, 1973; Matthews, 1995). Its effect has been to transform dissatisfaction and disobedience, as legitimate expressions of political freedom, into diseases.

Such criticisms of the disease model of insanity have been roundly rejected by the medical psychiatric establishment (Kendell, 1975b; Roth, 1976; Roth and Kroll, 1986). This is consistent with the scientific self-image of psychiatry. Yet behind the polemic, to which both sides in the debate about mental illness have been inclined, there are real difficulties. Whatever one's views about the concept of mental illness, there is no avoiding the fact that psychiatry is much trickier conceptually than physical medicine. Szasz's assimilation of hysteria to malingering may be too extreme: but the diagnosis of hysterical paralysis, say, is considerably more problematic, conceptually as well as practically, than that of its neurological counterparts. Similarly, it may seem too extreme to believe with Foucault that dissent, social or political, can never be pathological. But the diagnosis of mental illness is notoriously vulnerable to abuse, to being used as a means of political coercion, in ways and to an extent that that of physical illness is not (Bloch and Reddaway, 1977).

It is difficulties of this kind which, on a narrowly scientific medical model, locate psychiatry firmly as the rump end of scientific medicine. But it is also these same difficulties which have led to modern philosophical developments in this area, developments which as we will see have brought medicine and psychiatry closer together. We will return to these developments shortly after considering the relative neglect of these difficulties by bioethics.

IV. BIOETHICS AND THE NEGLECT OF PSYCHIATRY

Psychiatry is in many ways more problematic ethically than other areas of healthcare. It shares with such "high tech" disciplines as transplant surgery and molecular genetics, all the problems of consent, confidentiality, rationing, dual obligation and so forth, with which bioethics has traditionally been concerned. But to these it adds dimensions and depths of difficulty, which, though present in principle in more technologically based specialities, are writ large in the day-to-day of clinical work and research in psychiatry. It is only in psychiatry, for instance, that involuntary treatment of a fully conscious adult patient of normal intelligence in their own interests, is ethically justifiable; it is only in psychiatry, that not only treatment but diagnosis, prognosis, indeed the whole clinical process is routinely problematic ethically as well as scientifically; it is only in psychiatry that questions of value come overtly into the definitions of particular disorders, indeed into the very concept at the heart of the discipline, mental illness (see generally, Fulford, 1989 and 1993a).

Yet until recently, psychiatry has been relatively neglected by bioethics in favor of the more high profile problems of technological medicine. There are important exceptions: in an early volume in this series, Spicker and Engelhardt edited an impressive collection of articles about philosophical issues in mental health (Engelhardt and Spicker, 1978). More recently, Kopelman (1989), Hope (1990) and others have produced valuable reviews, and Bloch, Chodoff, and Green's (1999) edited collection is now in its third edition. But a glance through the back numbers of any bioethical journal shows the extent to which psychiatry has been largely eclipsed by other areas of healthcare.

The relative neglect of psychiatry by bioethics has extended even to the debate about mental illness. As we have seen, this debate actually turned in large part on whether mental illness is a moral or medical (i.e., scientific) concept. Yet although the debate has been vigorously pursued by psychiatrists, psychologists, sociologists, anthropologist, even by philosophers and historians (for a valuable collection see, Caplan, Engelhardt and McCartney, 1981), it has been widely ignored by bioethicists (with certain notable exceptions – see below). To a certain extent this was part and parcel of the neglect of more overtly ethical issues in psychiatry. Bioethics arose as a response to the ethical challenges posed by the increasing technological power of scientific medicine in the 1940s and 1950s. Hence psychiatry, and indeed primary care in general, although ethically more problematic, even to the point that their guiding concepts were openly evaluative in nature, were not within its immediate sights.

The neglect of psychiatry by bioethics, however, has also been a product of the predominantly scientific focus of psychiatry itself. As noted earlier, psychiatry has developed essentially by following in the scientific steps of physical medicine. Hence it has a considerable investment in the idea that mental illness, however value-laden it may appear, is, essentially, a purely scientific concept, just as (it is supposed) physical illness is purely scientific. Kendell (1975b), for example, in a strong statement of this view, concluded in defence of psychiatry that the two kinds of illness, mental and physical, occupy the same territory, a territory defined by objective norms of "biological dysfunction" (reduced life and/or reproductive potential). The anti-psychiatrists, of course, took exactly the same line but to the opposite conclusion. Szasz, in particular, emphasizing the ostensibly value-laden nature of mental illness, argued that mental illness was an *essentially* moral rather than medical (i.e., in his terms scientific) concept, defined by norms which are "ethical, legal and social in nature" (Szasz, 1960).

That bioethics, too, has often assumed an essentially medical model of mental illness, is graphically illustrated by a give-away turn of phrase which comes at the end of a detailed and careful account of rationality in the foundational *Principles* of Beauchamp and Childress (1989). Unlike many bioethics texts, this offers a thorough analysis of the links between loss of autonomy, incapacity and irrationality in the severely mentally ill; the analysis is developed in the context of a discussion of the key practical issue of involuntary psychiatric treatment; and there is a clear recognition that the relevant criteria of rationality are essentially value-laden. Yet far from concluding that psychiatric diagnosis, the basis for the ethical justification of involuntary psychiatric treatment, is essentially a value-laden procedure, Beauchamp and Childress argue that this makes the problem a "moral not a medical" one (Beauchamp and Childress, 1989, emphasis added). In other words, values, according to Beauchamp and Childress, and consistently with the medical model, are outside the scope of specifically medical (i.e., scientific) expertise.

As with the neglect of overtly ethical issues in psychiatry, there have been important exceptions to the neglect by bioethicists of the debate about mental illness, though in this case largely tangentially, among those, like Stuart Spicker, who have been concerned with the general philosophical problems lying behind bioethics. Thus, many wrote on the value-embedded nature of disease, attacking the claim that even physical illness is value-free (Agich, 1983; Engelhardt, 1975; Kopelman, 1990 and 1994); philosophical work on causality broadened the scope of bioethics to a more comprehensive philosophy of medicine (Nordenfelt and Lindahl, 1984; Spicker and Engelhardt, 1984); the importance of prognosis, too, was emphasized (Spicker and Raye, 1981); and as to conceptual issues generally, a wide range were tackled, notably in the *Philosophy and Medicine* series, including the core notions of disease and health (Nordenfelt, 1987).

It is important to recognize the extent to which even the philosophy of medicine, although concerned with a range of topics relevant to the debate about mental illness, has nonetheless remained relatively divorced from it. In writing of the value-embedded nature of disease, for example, there was little attempt to apply this to the apparent gap between the overtly value-laden mental illness and the apparently value-free physical illness (Fulford, 1989); in tackling causality in medicine (difficult enough, certainly), there were few attempts to extend this to the even more difficult problem of the relationship between causal and meaningful explanations, identified by no less a figure than Jaspers as being crucial to our understanding of the aetiology of severe mental illness (Jaspers, 1913b); prognosis was treated in statistical rather than intentional terms; and as to concepts, few explored in any detail the central conceptual significance of loss of autonomy (or incapacity) in marking out an experience *as* illness. Yet to be a patient, whether in psychiatry or physical medicine, is, exactly, as Stephen Toulmin (1980) pointed out, to be something less than a *full* agent.

This further underlines the extent to which the medical model has influenced not just the development of psychiatry, not just the debate about mental illness, not just traditional bioethics, but even the philosophy of medicine. In all these areas, the underlying though usually implicit guiding model has been of a core of scientific theory, for practical purposes free from evaluative considerations, by which the scope of properly medical expertise is defined, and to which ethical considerations are added as a gloss.

V. PHILOSOPHY, MEDICINE AND PSYCHIATRY

Given the success of science, it was to be expected that a medical scientific model should have dominated the development of psychiatry, not just in its scientific but also in its conceptual and bioethical aspects. After all, as Wittgenstein pointed out (Lee, 1980), the corresponding natural scientific model dominated three centuries of general philosophy! This model, though, is not as secure as it was. Even in physics, the simple mechanistic picture of the world assumed by nineteenth century science is no longer tenable. And in medicine, analytical philosophical work on the core concepts of disease, illness, health and disorder has started to spell out the many ways in which science itself, whether nineteenth- or twentieth-century in form, is not sufficient for a tenable model of healthcare.

A variety of new medical models are thus emerging. These differ in their focus and in their philosophical orientations. Some are linguistic-analytical (Fulford, 1989; Svenson, 1990), others existential (Wulff, Pederson and Rosenberg, 1968). Some emphasise health (Porn, 1984; Nordenfelt, 1987), others illness (Fulford, 1989), and still others disease (Svenson, 1990); they include equilibrium (Porn, 1984), potentiality (Nordenfelt, 1987), and biological (Svenson, 1990) and analytic (Fulford, 1989) accounts of the relationship between health, illness and disease. Many have relatively little to say about mental illness, assimilating it more or less readily to their analyses of physical illness (Porn, 1984; Nordenfelt, 1987). Others remain critical of the medical model of mental illness (Svenson, 1990). And yet others make mental illness central, arguing that it reveals features of the logic of medicine which are hidden by the apparently unproblematic nature of physical illness (Fulford, 1989).

In all these accounts, however, there is a recognition, in one form or another, and more or less tacitly, of the limitations of the conventional view that medicine is no more than a (descriptive) science. Where traditional medical models emphasize factual theories of disease developed in terms of disturbances of functioning, these more recent models emphasize the importance of the evaluative (as well as factual) elements in the meanings of health and disease, of the patient's experience of illness (as well as the doctor's knowledge of disease), and of this experience as involving an impairment of the agent's capacity to act (as well as of an organ's or system's capacity to function).

Each of these elements – values, the experience of illness, and loss of agency – has major implications for theory and practice in psychiatry. Thus, taking values first, value theory explains, straightforwardly, the more value-laden nature of mental illness compared with physical illness. This turns out to be no more than a product of a general logical feature of illness (as a value term) combined with the given diversity of peoples' values (Fulford, 1989, Ch. 5). Values, therefore, rather than being excluded by either the psychiatric or anti-psychiatric routes adopted in the debate about mental illness, are recognized to be endemic: they are woven into all aspects of healthcare, physical as well as mental; and they are significant not just in treatment choice (the focus of traditional bioethics), but in psychopathology, classification and diagnosis, aetiology and prognosis (Radden, 1994). This in turn has implications for the organization and delivery of services, for the role of "the patient's voice" in diagnosis as well as treatment (because of the significance of the patient's values); and for the role of the multi-disciplinary team especially in cross-cultural psychiatry (as providing a necessary rather than merely contingent balance of evaluative considerations (Fulford, 1995)). These implications are of proven practical importance, moreover. A neglect of the evaluative logical element in disease has directly contributed, for instance, to recent abuses of psychiatry (Fulford, Smirnoff and Snow, 1993) and to the development of unsatisfactory measures of the needs of the chronic mentally ill (Marshall, 1994).

The importance of the second element of the new medical models, the patient's experience of illness, is partly as it incorporates values. It goes beyond this, however, covering all aspects of the "lived experience" of illness, bodily as well as mental (Toombs, 1992). Again, the importance of this is in more than just treatment choice. The diagnostic significance even of the core symptoms of conditions such as schizophrenia can be shown to depend on the structure of beliefs and values within which they are set (Jackson and Fulford, 1997). With the classification of psychiatric disorders, similarly, the relevant conceptual elements include far more than the "pure" descriptions of the self-acclaimed "atheoretical" schemes currently available (see several articles in Sadler, Wiggins and Schwartz, 1994). A merely descriptive psychopathology is too narrow, then. It is essential, but not sufficient. There is a need, not to overthrow, but to complete Kraepelin's system (Fulford, 1998). With Kraepelin we focused on symptoms. What is now required is attention to those features of the experience of illness which mark out symptoms – pains, anxieties, appetites, compulsion, beliefs or whatever – as symptoms (Fulford, 1989, Ch. 7). Such a move will require the resources not only of Anglo-American but also of Continental Philosophy (Graham and Stephens, 1994; Sadler, Wiggins and Schwartz, 1994). This is as it should be. Scientific psychiatry itself grew out of a period when the two great philosophical traditions were less divorced. And the pressure for a reconciliation at the present time, the pressure for better ways of describing subjective experience and conceptualizing its links with brain activity, is coming most strongly from developments at the cutting edge of neuroscience, in dynamic brain imaging and artificial intelligence (Spitzer, 1995).

The epicenter of this renewed interest in subjective experience, however, in neuroscience as well as philosophy, is the third conceptual element emphasized by the new medical models, the concept of agency. As Boden (1990) has put it, the neurosciences are taking a renewed interest in the "warm blooded" mental activities – conation, desire, volition – in contrast to the merely cognitive. This coincides precisely with the importance of loss of agency as the key feature of the experience of illness. The elements of what Austin (1956/57) called the "machinery of action" are explicit throughout classifications of mental disorders (Fulford, 1994) and they are crucial to the core symptom of delusion (Fulford, 1993b). Recognizing this allows us to draw on a richer set of conceptual tools for analyzing psychopathology. This in turn is critical to diagnosis, especially in areas such as forensic psychiatry (mad or bad) and involuntary treatment (mad or sad), involving as they do issues of responsibility and the assessment of "capacity." It also generates new models of the aetiology of mental disorders, reconciling cause and meaning, and suggesting improvements in the management of such conditions as post-traumatic stress disorder (Bolton and Hill, 1996). This reconciliation, furthermore, reflects a broader reconciliation of mind and body. The distinct mental and bodily functions identified in the traditional medical model reinforce a Cartesian picture of medicine; but agency, or human agency at any rate, requires an embodied mind.

Given the reconciliation of mind and body implicit in the notion of agency, it is natural that we should find the elements of the new philosophy of psychiatry turning up in areas of the philosophy of bodily medicine. I have already noted the work of a number of authors on the evaluative nature of the medical concepts; and the recent widespread dissatisfaction with the traditional medical model of disease. But the many strands of the new philosophy are all drawn together to a remarkable degree in the work of Stuart Spicker. As a bioethicist, Stuart Spicker has been no slouch. He has written incisively on most of the major issues as they have become current – consent, the market in health care, bioethics education, and so forth. He has repeatedly emphasized the pervasiveness of values (e.g., Engelhardt and Spicker, 1975); and the importance of healthcare professions other than medicine (Spicker and Gadow, 1980). Innovative and tireless, he has helped to spear-head several major publishing ventures, responding to issues of the day. His latest, *Healthcare Ethics Forum*, is proving a valuable resource for lay and professional ethicists alike on both sides of the Atlantic.

In addition to all this, however, he has had a keener eye than most to psychiatry, and to the deep metaphysical problems it raises. Influenced as he was by a philosophically-minded psychiatrist, Erwin W. Strauss (Spicker, 1978), he was well placed to develop this side of his work (Engelhardt and Spicker, 1976 and 1978). Especially in his early publications, this bore important fruits, not least a highly pertinent rejection of the Cartesian "psyche" (Spicker, 1970a; 1973a; 1976); and a continuing ability to draw equally on both sides of the Continental/Analytical philosophical divide (e.g., Spicker, 1987). Most of his more recent research has been less obviously connected with psychiatry. But each of the three key elements of modern work on the concept of mental disorder are clearly present as continuing themes in his work in the philosophy of medicine: the pervasiveness of values (for instance, in diagnosis, Spicker, 1991, and in prognosis, Spicker and Raye, 1981); valuable insights into both the structure of consciousness (Spicker, 1973a) and the "lived body" (Spicker, 1975), which indeed reflect his own important early studies in phenomenology (Spicker, 1970b and 1973b); and a recognition of the significance of aspects of agency other than mere cognition (e.g., in Spicker, 1986).

In Stuart Spicker's wide ranging work in the philosophy of medicine there is thus a remarkable point-for-point harmony with developments in the philosophy of psychiatry. Mind and body, mental and physical, are in step. As twin attributes of persons, so they should be (Strawson, 1977). But it is the effective denial of this by medicine, relegating the mind to second place, which is at the heart of modern objections to the traditional medical model. And it is the affirmation of both mind and body, on an equal basis, which is at the heart of the new models of medicine now emerging from philosophical work in this area.

VI. CONCLUSIONS AND FUTURE DIRECTIONS

As the philosophy of medicine and the philosophy of psychiatry converge, so they point to a future in which diversity, disease and determinism are each perceived quite differently from the way they have been perceived in the traditional medical model. As we saw earlier, in the traditional model the diversity of psychiatry's theories was taken to be a mark of its scientific underdevelopment (it is a pre scientific discipline, some wanted say, e.g., Boorse, 1976); the concept of mental illness, similarly, stood (for psychiatrists) or fell (for anti-psychiatrists) according to whether it passed the test of a deterministic disease model like the biological disease models of physical illness; and because psychiatry, even in its own self-image, was a junior scientific partner of physical medicine, so its ethical problems were largely subsumed to those of "high-tech" medicine by much of bioethics.

In the philosophy of medicine now evolving, however, all this will be reversed. Just how things will develop, only time will tell. But it seems likely, first, that the division (antagonism?) between mind and body in medicine will be less clear cut. This will be important practically; psychosomatic disorder (combining mental and bodily, elements) will increasingly be the norm. But it will also be important for theory. As Stuart Spicker's early work implied (Spicker, 1976), there is a spuriousness about the separation of psyche and soma which has prejudiced the development of physical medicine as much as psychiatry. In so far, therefore, as a diversity of theoretical models remains appropriate for psychiatry, so it will become increasingly appropriate for the rest of medicine. Far from psychiatry following physical medicine, then, and becoming less diverse in its models, physical medicine will follow psychiatry, and become more diverse!

A similar role reversal seems likely in relation to disease theory. The difficulty psychiatry has found in following physical medicine in this respect has traditionally been perceived as a failure of psychiatric science: as we saw, psychiatry's critics have taken this as evidence that mental illness does not exist. But the science involved in, say, liver disease, is not necessarily an appropriate model for diseases of the brain. In particular, the kind of determinism exhibited by the well-behaved mechanisms studied by nineteenth century mechanics, and which has been largely adopted in the medical model, is unlikely to be exhibited by the most complex mechanism we have yet encountered, the brain. It seems overwhelmingly likely that brains will be, literally, disorderly; like the weather, such patterns as they display, will be the patterns of chaotic systems, so sensitive to their initial conditions that determinate predictions are impossible even in principle. More contentiously, but also more intriguingly, they may share with quantum mechanics some of the deeper indeterminacy introduced when the observer cannot, again even in principle, be wholly disentangled from the observed (Fulford, 1996).

To the extent, then, that anything resembling a disease model of mental illness proves possible, it will not be deterministic in the way that the simpler disease models appropriate to physical medicine are deterministic. Here too, though, psychiatry will be leading the way. For, to the extent that the psyche/soma divide breaks down in medicine, so to the same extent the more complex disease models (putatively) appropriate to the psyche will become appropriate in medicine more generally. The point is, perhaps, a general one, not limited to mind and brain. The traditional medical model works well with isolated systems. But disease processes are increasingly being understood ecologically: it is a system's environment, not (merely) its isolated internal functioning, which determines its state of health or disease. And the more complex the arrangements of systems that we have to consider, the less likely are they to behave in a well ordered, mechanistic, way.

All this begs a key question, of course. For the notion of agency, introduced here at the heart of recent analyses of illness, brings with it more than a merely negative lack of determinate predictability. Agents – human agents at any rate, or persons – also have a positive freedom of action. It is this which disease, on whatever model of determinism, impairs. A person who is ill is less self-determined (capable of determining their own actions) and more determined (by their disease).

Modern psychiatric treatments aim to restore agency, to give us back control. The freedom gained is, perhaps, a psychological freedom only: for metaphysical freedom, something stronger than medicine is called for! But it is a freedom worth having, nonetheless, and a freedom to which the many strands of Stuart Spicker's work unerringly point.

University of Warwick, and
University of Oxford
United Kingdom

BIBLIOGRAPHY

Agich, G. J.: 1983, 'Disease and Value: A Rejection of the Value Neutrality Thesis', *Theoretical Medicine* 4, 27–41 .

Austin, J. L. : 1956–57, 'A Plea for Excuses', *Proceedings of the Aristotelian Society* 57, 1–30. Reprinted in White, A. R. (ed.): 1968, *The Philosophy of Action*, Oxford, Oxford University Press.

Beauchamp, T. L. and Childress, J. F.: 1989, *Principles of Biomedical Ethics* (third edition), Oxford, Oxford University Press.

Bloch, S., Chodoff, P., and Green, S.: 1999, *Psychiatric Ethics* (third edition), Oxford, Oxford University Press.

Bloch, S. and Reddaway, P.: 1977, *Russia's Political Hospitals: The Abuse of Psychiatry in the Soviet Union*, Southampton, The Camelot Press.

Boden, M. A.: 1990, *The Creative Mind: Myths and Mechanisms*, London, Butler and Tanner Ltd.

Bolton, D. and Hill, J.: 1996, *Mind, Meaning and Mental Disorder: The Nature of Causal Explanation in Psychology and Psychiatry*, Oxford, Oxford University Press.

Boorse, C.: 1976, 'What a Theory of Mental Health Should Be', *Journal of Theory Social Behaviour* 6, 61–84.

Caplan, A. L., Engelhardt, H. T., and McCartney, J. J. (eds.): 1981, *Concepts of Health and Disease: Interdisciplinary Perspectives*, Addison-Wesley Publishing Co., Reading, Massachusetts.

Christen, Y. and Churchland, P: 1992, *Neurophilosophy and Alzheimer's Disease*, Springer-Verlag, Heidelberg.

Clare, A.: 1979, 'The Disease Concept in Psychiatry', Ch 3 in P. Hill, R. Murray, and A. Thorley (eds.), *Essentials of Postgraduate Psychiatry*, Academic Press, Grune and Stratton, New York.

Engelhardt, H. T. Jr. : 1975, 'The Concepts of Health and Disease', in H. T. Engelhardt Jr. and S. F. Spicker (eds.), *Evaluation and Explanation in the Biological Sciences*, D. Reidel, Dordrecht, Holland.

Engelhardt, H. T., Jr., and Spicker, S. F. (eds.): 1975, *Evaluation and Explanation in the Biomedical Sciences*, D. Reidel Publishing, Dordrecht, Holland.

Engelhardt, H. T., Jr., and Spicker, S. F. (eds.): 1976, *Philosophical Dimensions of the Neuro-Medical Sciences*, D. Reidel Publishing, Dordrecht, Holland.

Engelhardt, H. T., Jr., and Spicker, S. F. (eds.): 1978, *Mental Health: Philosophical Perspectives*, D. Reidel Publishing, Dordrecht, Holland.

Foucault, M.: 1973, *Madness and Civilization: A History of Insanity in the Age of Reason*, Random House, New York.

Fulford K. W. M.: 1989, reprinted 1995, *Moral Theory and Medical Practice*, Cambridge University Press, Cambridge.

Fulford, K. W. M.: 1993a, 'Bioethical Blind Spots: Four Flaws in the Field of View of Traditional Bioethics', *Health Care Analysis* 1, 155–162.

Fulford, K. W. M.: 1993b, 'Mental Illness and Mind-brain Problem: Delusion, Belief and Searle's Theory of Intentionality', *Theoretical Medicine* 14, 181–194.

Fulford, K. W. M.: 1994, 'Closet Logics: Hidden Conceptual Elements in the DSM and ICD Classifications of Mental Disorders', in J. Z. Sadler, O. P. Wiggins, and M. A. Schwartz (eds.), *Philosophical Perspectives on Psychiatric Diagnostic Classification*, ch. 9, Johns Hopkins University Press, Baltimore.

Fulford, K. W. M.: 1995, 'Concepts of Disease and the Meaning of Patient-centred Care', in K. W. M. Fulford, S. Ersser, and T. Hope (eds.), *Essential Practice in Patient-Centred Care*, Ch. 1, Blackwell Science, Oxford.

Fulford, K. W. M.: 1996, 'Religion and Psychiatry: Extending the Limits of Tolerance', in D. Bhugra (ed.), Routledge and Kegan Paul, London, Ch. 1.

Fulford, K. W. M.: 1998, 'Completing Kraepelin's Psychopathology: Insight, Delusion and the Phenomenology of Illness', in X. F. Amador and A. S. David (eds.), *Insight and Psychosis*, ch. 2, Oxford University Press, New York.

Fulford, K. W. M., Smirnoff, A. Y. U., and Snow, E.: 1993, 'Concepts of Disease and the Abuse of Psychiatry in the USSR', *British Journal of Psychiatry* 162, 801–810.

Graham, G., and Stephens, G. Lynn. (eds.): 1994, *Philosophical Psychopathology*, MIT Press, Cambridge, Massachusetts.

Hope, R. A.: 1990, 'Ethical Philosophy as Applied to Psychiatry', *Current Opinion in Psychiatry* 3, 673–676.

Jackson, M., and Fulford, K. W. M.: 1997, 'Spiritual Experience and Psychopathology', *Philosophy, Psychiatry, and Psychology* 4(1), 41–66.

Jaspers, K.: 1913a, *Allgemeine Psychopathology*, Springer Verlag, Berlin. English edition, Hoenig, J. and Hamilton, M. W.: 1963, *General Psychopathology*, Manchester University Press, Manchester.

Jaspers, K.: 1913b, 'Causal and Meaningful Connexions between Life History and Psychosis', in S. R. Hirsch and M. Shepherd (eds.), 1974, *Themes and Variations in European Psychiatry*, ch. 5, John Wright and Sons Ltd., Bristol.

Kendell, R. E.: 1975a, *The Role of Diagnosis in Psychiatry*, Blackwell Scientific Publications, Oxford.

Kendell, R. E.: 1975b, 'The Concept of Disease and its Implications for Psychiatry', *British Journal of Psychiatry* 127, 305–315.

Kenny, A. J. P.: 1969, 'Mental Health in Plato's Republic', *Proceedings of the British Academy* 5, 229–253.

Kopelman, L. M.: 1989, 'Moral problems in Psychiatry', in R. Veatch (ed.), *Medical Ethics*, Jones and Bartlett Publishing Company, Boston.

Kopelman, L. M.: 1990, 'On the Evaluative Nature of Competency and Capacity Judgements', *International Journal of Law and Psychiatry* 13, 309–329.

Kopelman, L. M.: 1994, 'Normal Grief: Good or Bad? Health or Disease?', *Philosophy, Psychiatry, and Psychology* 1, 209–220. (Commentaries by Dominian, J. and Wise, T.N. pp. 221–224; response by Kopelman, pp. 226–227.)

Lee, D. (ed.): 1980, *Wittgenstein's Lectures: Cambridge, 1930–1932*, Basil Blackwell, Oxford.

Marshall, M.: 1994, 'How Should We Measure Need? Concept and Practice in the Development of a Standardised Assessment Schedule', *Philosophy, Psychiatry, and Psychology* 1, 27–36. (Commentaries by Crisp, R and Morgan, J., pp. 37–40.)

Matthews, E.: 1995, 'Moralist or Therapist? Foucault and the Critique of Psychiatry', *Philosophy, Psychiatry, and Psychology* 2, 19–30.

Nordenfelt, L.: 1987, *On the Nature of Health: An Action-Theoretic Approach*, D. Reidel Publishing Company, Dordrecht.

Nordenfelt, L. and Lindahl, B. I. B. (eds.): 1984, *Health, Disease, and Causal Explanations in Medicine*, D. Reidel Publishing Company, Dordrecht.

Parsons, T.: 1951, *The Social System*, Free Press, Glencoe, Illinois.

Porn, I.: 1984, 'An Equilibrium Model of Health', in L. Nordenfelt and B. I. B. Lindahl (eds.), 1984, *Health, Disease, and Causal Explanations in Medicine*, D. Reidel Publishing Company, Dordrecht.

Radden, J.: 1994, 'Recent Criticisms of Psychiatric Nosology: A Review', *Philosophy, Psychiatry and Psychology* 1, 193–200.

Roth, M.: 1976, 'Schizophrenia and the Theories of Thomas Szasz', *British Journal of Psychiatry* 129, 317–326.

Roth, M. and Kroll, J.: 1986, *The Reality of Mental Illness*, Cambridge University Press, Cambridge.

Sadler, J. Z., Wiggins, O. P., and Schwartz, M. A.: 1994, *Philosophical Perspectives on Psychiatric Diagnostic Classification*, The Johns Hopkins University Press, Baltimore.

Spicker, S. F.: 1970a, *The Philosophy of the Body: Rejections of Cartesian Dualism*, Quadrangle Books, Inc., Chicago. (Reprinted: New York Times Book Company, 1973.)

Spicker, S. F.: 1970b, 'Shadworth Hodgson's Reduction as an Anticipation of Husserl's Phenomenological Psychology', *Journal of the British Society of Phenomenologists* 2, 57–73.

Spicker, S. F.: 1973a, 'The Fundamental Constituents of Consciousness: Process-contents and the Erlebnisstrom', *Man and World* 6, 26–43.

Spicker, S. F.: 1973b, 'Inner Time and Lived-through Time: Husserl and Merleau-Ponty', *Journal of the British Society of Phenomenologists* 4, 325–347.

Spicker, S. F.: 1975, 'The Lived-body as Catalytic Agent: Reaction at the Interface of Medicine and Philosophy', in H. T. Engelhardt Jr. and S. F. Spicker (eds.), *Evaluation and Explanation in the Biomedical Sciences*, D. Reidel Publishing Company, Dordrecht, pp. 181–204.

Spicker, S. F.: 1976, 'The Spurious Psyche – Some Distinctions', in H. T. Engelhardt Jr. and S. F. Spicker (eds.), *Philosophical Dimensions of the Neuro-Medical Sciences*, D. Reidel Publishing Company, Dordrecht, pp. 163–177.

Spicker, S. F.: 1978, 'The Psychiatrist as Philosopher: Gewidmet Erwin W. Straus', (October 11, 1891–May 20, 1975), in H. T. Engelhardt Jr. and S. F. Spicker (eds.), *Mental Health: Philosophical Perspectives*, D. Reidel Publishing Company, Dordrecht pp. 144–155.

Spicker, S. F.: 1986, 'Cognitive and Conative Issues in Contemporary Philosophy of Medicine', *Journal of Medicine and Philosophy* 11, 107–117. Translated into Chinese and reprinted in Zhexue Yicong [Philosophical Translations], Vol. 3, 1987, pp. 73–78.

Spicker, S. F.: 1987, 'An Introduction to the Medical Epistemology of George Canguilhem: Moving beyond Michel Foucault', *Journal of Medicine and Philosophy* 12, 397–414.

Spicker, S. F.: 1991, 'Ethics in Diagnosis: Bodily Integrity, Trust-telling, and the Good Physician', in J. L. Peset and D. Gracia (eds.), *The Ethics of Diagnosis*, Kluwer Academic Publishing, Dordrecht., pp. 107–132.

Spicker, S. F. and Engelhardt, H. T.: 1984, 'Causes, Effects and Side Effects', in L. Nordenfelt and I. B. Lindahl (eds.), *Health, Disease and Causal Explanations in Medicine*, D. Reidel Publishing Company, Dordrecht, pp. 224–233.

Spicker, S. F., and Gadow, S.: 1980, *Nursing: Images and Ideals (Opening Dialogue with the Humanities)*, Springer Publishing Company, New York.

Spicker, S. F. and Raye, J. R.: 1981, 'The Bearing of Prognosis on the Ethics of Medicine: Congenital Anomalies, the Social Context and the Law', in S. F. Spicker, J. M. Healey, and H.T. Engelhardt Jr. (eds.), *The Law-Medicine Relation: A Philosophical Exploration*, D. Reidel Publishing Company, Dordrecht, pp. 189–216.

Spitzer, M.: 1995, 'Conceptual Developments in the Neurosciences Relevant to Psychiatry', *Current Opinion in Psychiatry* 8, 317–329

Storr, A.: 1989, *Freud*, Oxford University Press, Oxford.

Strawson, P. F.: 1977, *Individuals: An Essay in Descriptive Metaphysics*, Oxford University Press, Oxford.

Svenson, T.: 1990, *On the Notion of Mental Illness*, Department of Health and Society, Linkoping, Sweden.

Szasz, T. S.: 1960, 'The Myth of Mental Illness', *American Psychologist* 15, 113–118.

Szasz, T. S.: 1987, *Insanity: The Idea and its Consequences*, John Wiley and Sons, New York.

Toombs, S. K.: 1992, *The Meaning of Illness: A Phenomenological Account of the Different Perspectives of Physician and Patient*, Kluwer Academic Publishing, Dordrecht.

Toulmin, S.: 1980, 'Agent and Patient in Psychiatry', *International Journal of Law and Psychiatry* 3, 267–278.

Wing, J. K.: 1978, *Reasoning about Madness*, Oxford University Press, Oxford.

Wulff, H. R., Pedersen, S. A. and Rosenberg, R.: 1968, *Philosophy of Medicine, an Introduction*, Blackwell Publications, Oxford.

BERNARD M. DICKENS

EQUITY AND THE HEALTH EFFECTS OF URBANIZATION

I. INTRODUCTION

One of the many pleasures of contributing to a *festschrift* for Stuart Spicker is the wide scope available to address a topic of one's choice in the confidence that it will relate to some aspect of Stuart's professional career and accomplishments. The aspect that provides the hook on which this essay is hung is Stuart's status as a bioethicist of international stature. Following completion of doctoral studies in the U.S., Stuart held a Post-Graduate Research Fellowship of the National Endowment for the Humanities in Cambridge, England and launched a career that has included visiting professorships in Germany, France, England and Argentina, consultancies in Israel and the People's Republic of China, awards in Japan and extensive publication in Japan and particularly from the Netherlands, while at the same time remaining intensively involved in the development of bioethical teaching, lecturing and writing in the U.S. Invited presentations made not only throughout the U.S. but also in many cities in England and Wales, Germany, Sweden, Israel, Canada, France, the Netherlands, Italy, Hungary, Malta and in innumerable cities of Japan have established Stuart as a scholar of truly international reputation.

My own friendship with Stuart grew out of the inaugural meeting of international colleagues in a variety of disciplines related to bioethics at Milazzo, Sicily in 1989, after which my wife and I had the pleasure of traveling on the island with Stuart and his wife, Lorraine, whose Canadian origins I appreciated. I observed Stuart's subsequent career activity with growing admiration, and noted his significant contributions to bioethics not only in the U.S. and Canada, but also, for instance, in Germany, at Oxford University in England, and particularly in lecture tours of Japan. The esteem in which Stuart is held in Japan provides the link to this essay in two regards. The essay is based on perceptions developed in the context of the opening symposium of the World Health Organization's Centre for Health Development in Kobe, Japan, in March 1996 on the topic of "Urbanization – A Global Health Challenge," and

H. T. Engelhardt, Jr. (ed.), The Philosophy of Medicine, 177–194.

Tokyo is one of the world's increasingly numerous "megacities," meaning cities whose populations exceed ten million people. The effects on individual and communal health of urbanization, and the pursuit of equity in clinical and public health care of urban communities are the focus of the essay.

II. URBANIZATION

The appearance of vast cities, through immigration of population from rural areas and other countries, and indigenous growth through birth rate escalation and adult longevity, is a phenomenon of the present time. It is estimated that while in 1994 about 45% of the world's population lived in urban areas, by the year 2025 about 61% of people will be urban dwellers, and in Latin America and North America the figure will be 85% (Rees, 1996). Within 15 years, between 20 and 30 cities are predicted to have populations of over 20 million (Goldstein and Kickbusch, 1996), and all but a few such as Tokyo, New York and Los Angeles will be in developing countries. Rural land between cities will vanish as cities merge into conurbations with nothing but street signs and similar features of urban topography to mark their boundaries. Tokyo has already become fused with the historically distant port city of Yokohama, and it was observed for instance at the Kobe symposium that, in the U.S., urban development is almost uninterrupted linking Boston, Providence and New Haven through New York, New Jersey, Philadelphia and Baltimore to Washington, D.C. Green space between is occupied by suburban residential developments, industrial parks and landscaped commercial structures and shopping malls, rather than by land in agricultural use. In Asia, it is predicted that rapid economic growth and the rise in urban populations will combine to double the number of megacities to 17 or 18 by the year 2020, and that some centres such as Tokyo, Shanghai and Bombay will have populations approaching 30 million people ('Managing Asia's Cities,' 1996, p. 23), each exceeding the total population of Canada in 1996.

Economic growth may allow cities such as Tokyo to provide adequate clinical health services and to enjoy some success in overcoming the hazards to community health associated with solid and human waste, air pollution from industrial and motor vehicle emissions and, for instance, contamination by industrial effluents and human waste of water used for drinking, domestic purposes and crop irrigation. Urban populations of poorer countries are vulnerable, however, to air, water and ground contaminants and the health hazards of living in underserviced, overcrowded tenement blocks or in squatter or shanty settlements, often located on marginal land exposed to flooding, industrial smog or landslides. Human factors of family breakdown, poverty, violence, social marginalization and substance addiction often aggravate ill health. Noting that by 1990, at least 600 million people in the world's urban areas of developing countries were already living in life-threatening and health-threatening conditions, W.H.O. officers have observed that

> People in cities – particularly the poor and the newly arrived – experience stresses and exposures that result in a wide range of health problems, including communicable diseases, malnutrition, mental illnesses and chronic respiratory diseases. Unhealthy conditions include poverty, inadequate food and shelter, insecure tenure, physical crowding, poor waste disposal, unsafe working conditions, inadequate local government services, overuse of harmful substances and environmental pollution (Goldstein and Kickbusch, 1996).

These are the conditions that policy-makers and resource managers face in their task of achieving health care equity.

III. EQUITY

The concept of equity is understood in several different but related senses. It is usually taken to mean fairness or justice, which is the ethical principle to which the law is devoted. It may be taken to mean equality, contributing to the principle of justice that like cases, circumstances and people should be treated alike. This principle also suggests, however, that different cases should be treated differently, on the ground that it is inequitable or unfair to treat different cases, circumstances or groups of people alike when there are relevant or significant differences between them. The principle is not that these different people should be treated equally, as if they were the same, but that they should all be treated fairly in light of their differences. This raises the question of the senses in which populations may be considered like each other, and different from each other; that is, the ways in which population groups may be compared and contrasted. People may appear different in some regards, related, for instance, to whether they reside in urban or rural areas, but the same in significant other regards, such as their equitable entitlement to protection against the various hazards to their health that they experience. They may also appear the same, in that, for instance, they reside in urban areas, but significantly different in that, for instance, some have access to good quality health care while others inequitably do not, perhaps because they are poor or powerless.

Many publicly funded health care programs operate on the principle that everyone eligible for coverage (on such bases as citizenship, residence or age) should be assured of reasonable access to a set of necessary medical services. Eligibility criteria may exclude illegal immigrants and residents, however, and deny services to homeless people who cannot show that they live within a relevant catchment area for services. Medical and related care will be provided principally without direct charge to eligible recipients, on the basis of their assessed medical needs. Services may be available through public hospitals, local clinics and, for instance, community-based nursing centers, which attempt to provide an appropriate level of care and access within proper time without undue delay or inconvenience. Once equity is assured in patients' reasonable access to essential health care services, persons with the means may avail themselves of additional services, exceeding or extending beyond those provided under the program.

The principle of equity is satisfied in that eligible patients are assured of the essential services they are shown to need without regard to their means to pay, and is not violated if those who are able to pay for additional care choose to obtain it. The result may be inequality of care, but not inequity. Fairness is satisfied when everyone's basic medical needs are provided for, and not offended when people with the means to provide themselves with care beyond their basic needs want to use their resources on health-related services rather than on alternative luxuries or indulgences. The residual inequity concerns the narrowness of eligibility criteria when they exclude coverage for instance of illegal squatter residents, whose health care needs are liable to be profound.

Comparisons and contrasts in health and health care may ethically be made at different levels. Modern ethics in western countries have tended to focus on the entitlements and duties of individuals, and ethics that address health and human biology, often described as bioethics, promote the value of individual autonomy and self-determination. Ethical orientations and principles applied to relations between two or more individuals operate at the micro-ethical level. Historically, however, ethics has also been at work among different social or population groups, and between such groups and individual persons, including group members and those considered to be outside a group. This level of ethical assessment is described as the macro-ethical level, where macro-ethics apply. It has been observed that:

> Macro-ethics can be defined as a set of principles designed to protect the human dignity, integrity, self-determination, confidentiality, rights and health of populations and the people comprising them. The kinds of social grouping encompassed in this definition include communities, cultures, social orders, and populations of various descriptions including racial, ethnic, religious, cultural and other minorities (Gostin, 1991, p. 30).

Issues of equity in urbanization concern comparisons and contrasts among subgroups of an urban population, relations among populations of different urban centers, between central urban and suburban populations and, for instance, between urban and rural populations. They raise discussion at the macro-ethical level rather than at the level of relations between and among particular individual persons.

Challenges to equity in health and health care presented by urbanization therefore raise comparisons and contrasts in health status and in access to health care services between urban and rural populations of a country or region of a country, between urban inner-city and suburban populations and, for instance, among different socio-economic groups or classes within an urban community. Transcending national boundaries of macro-ethics are wider equitable comparisons and contrasts among urban populations of different countries in the same region, continent or hemisphere, and among those in different regions, continents or hemispheres, such as north and south. Wider comparisons and contrasts may engage the attention of international agencies and institutions more effectively than that of individual nations and cities.

Equity may entail equality or fairness in health status or treatment outcomes for comparable or different populations, equality or fairness in health opportunities for populations or, for instance, in access to good quality health care services. What constitutes good quality services also warrants comparison and contrast. Some populations, notably in western countries, may make comparative assessments by reference to use of advanced medical technologies and access to drugs and vaccines. Surgical interventions and use, for instance, of sophisticated equipment may furnish the components of "good" care.

In contrast, others may prefer comparisons based on home or community care rather than institutional care, and measure the quality of treatment by reference to the number of personnel actively engaged in caring for a sick person, or the time that individual care-givers devote to interacting personally with patients and affording them support and comfort, meeting their medical needs as far as resources allow. The World Health Organization recognizes that health includes more than medical care; its Constitution describes "health" as a state of physical, mental and social well-being, and not merely the absence of disease or infirmity. Accordingly, populations that understand and require good quality health care by reference to different criteria may frustrate meaningful equitable comparison and contrast. Since they are different in this material regard, caution must be exercised in assessing their health by uniform measures.

As against this, however, urban, suburban and rural populations and subdivisions of such populations, however different in their compositions and cultures, are similar in some central expectations. They intend, for instance, that women will survive pregnancy and childbirth, that infants will survive to become children, that children will live to become adolescents, that adolescents will reach mature adult years and that adults will be able to found families at the time and of the size they want. Therefore, relevant comparisons and contrasts may be drawn of maternal mortality and morbidity, of infant and child survival and health and, for instance, of infertility and planned and unplanned reproduction.

It is an inescapable matter of ethical concern to observe the contrast that:

> In the industrialized countries maternal deaths are now rare: the average life time risk for a woman dying of pregnancy-related causes is between 1 in 4,000 and 1 in 10,000. For a woman in the developing countries, the average risk is between 1 in 15 and 1 in 50. These countries have maternal mortality rates 200 times higher than those of Europe and North America (Mahler, 1987).

However, racial differences can exist within countries too. In the U.S. in 1986, for instance, maternal mortality for white women was 4.9 deaths per 100,000 live births, but the death rate for black women was 15 per 100,000 live births (National Centre for Health Statistics, 1988). Accordingly, notwithstanding refinements of cultures and circumstances among various populations, health assessments can be proposed by which to make equitable comparisons and contrasts, and to identify inequities.

IV. ENVIRONMENTAL EQUITY

It is commonly observed that good health and access to good health care services are related to such factors as high income, comfortable housing, literacy and advanced education, rewarding work in a salutary workplace, leisure and recreational opportunities, nutritional and balanced diet and, for instance, social and political empowerment. Accumulated research indicates that "the most important determinant of health – in the United States, Canada, and every other country – is the standard of living. Numerous peer-reviewed medical studies reach the same conclusion" (see Cross, 1995, p. 753, and references cited in notes 114–115). Good standards of living are not equitably available in most urban settings. Favored urban communities are frequently able to protect their residential neighborhoods against such environmental risks to health as being locations of toxic and otherwise hazardous waste sites, polluting or abandoned industrial plants, contaminated water supplies and foodstuffs and, for instance, municipal landfills and garbage incinerators.

In contrast, underprivileged urban areas are likely to be afflicted not only with substandard housing and the manifold consequences of social deprivation, stress and disorder, but with an accumulation of environmental hazards to health. The economic and political powerlessness of their populations leaves them defenseless against planning decisions and unplanned neglect of sound environmental practices that condemn them to live in pollution and contamination. Shanty-town residents are frequently unenfranchised in muncipal government, and powerless to contribute to governmental planning decisions that affect them. In many multiracial countries, urban populations at risk are so evidently of a minority or disempowered racial identity as to cause their geographical health disadvantage to be described as "environmental racism" (Wigley and Shrader-Frechette, 1996).

When racial and other disadvantaged urban populations lack political power, danger to their health standards is liable to be aggravated by an accumulation of municipal and wider governmental assessments that undervalue risks to health from decisions to locate or leave environmental threats among them. Empowered populations successfully resist such adverse assessments and decisions by so-called "NIMBY" (Not In My Back-Yard) campaigns to direct sources of threat to health elsewhere. Strong and frequently courageous governmental leadership is required to achieve urban environmental equity. In the U.S. on Earth Day, 1993, President Clinton announced a federal action plan to achieve environmental justice. On February 11, 1994, he signed an Executive Order directing each U.S. federal agency to develop an environmental justice strategy for "identifying and addressing . . . disproportionately high and adverse human health or environmental effects of its programs, policies, and activities on minority populations and low-income populations (Wigley and Shrader-Frechette, 1996, p. 57). It is a matter of international interest to see how successful such strategies will prove to be.

Beyond health hazards from the physical environment of low-income areas, hazards to health in the full sense in which the World Health Organization understands the concept, embracing physical, mental and social well-being, may arise in such areas from the high incidence of overcrowding, violence, gang-related and drug-related crime, including drug-related burglaries, thefts and robberies. Where urban policing services are funded on a local basis, low-income and economically disadvantaged areas tend to lack the tax base to furnish well staffed, trained and equipped police forces. Local law-abiding residents lack protections against the depredations of others who can defy the law with relative impunity. Property offences and domestic abuse no doubt occur in more affluent areas too, but there, non-domestic offences are more liable to take the form of commercial frauds and vehicle and similar thefts against the material costs of which affluent populations can achieve commercial insurance as well as police protection. Feelings of physical and property insecurity can be felt in affluent urban areas, but the reality of danger, injury and loss represent a more pervasive and direct hazard to health in deprived districts of cities.

V. EQUITABLE HEALTH RESOURCE ALLOCATIONS

In addressing social justice and equal access to health care services, and principles by which scarce health resources may be allocated, the ethicist Gene Outka has asked, "Which then among the standard conceptions of social justice appear to be particularly relevant or irrelevant?" (Powers, 1996, p. 81). He considers five options applicable to allocation of resources among individuals, which may be expanded to apply to urban population sub-groups. These are:

1. To each according to its merit or desert.
2. To each according to its societal contribution.
3. To each according to its contribution in satisfying whatever is freely desired by others in the open marketplace of supply and demand.
4. To each according to its needs.
5. Similar treatment for similar cases.

The first relies on possibly invidious distinctions being drawn in order to rank urban population sub-groups on a vertical scale that, at its lower levels, identifies communities found less than others to merit or deserve health protection or health care. Even if distinctions of this nature could be made by a disinterested agency or process that maintained public trust, the potential of this principle to result in indefensible denials of health care, and to compel those denied allocations of resources to subsidize care given to groups that receive higher assessments, would leave the principle difficult to support on grounds of equity. Nevertheless, where "environmental racism" and its analogies occur it may be concluded that populations whose health is jeopardized are being assessed negatively according to this principle, whether deliberately or by implication.

The second allocation principle, societal contribution, can be equally objectionable. Where wealthy populations are proportionately taxed, they obviously contribute more to a society's material resources than poorer populations. In so far as more wealthy populations tend to have better health status and health care services, and reside in more congenial neighborhoods, it may appear that health resource allocation is indeed organized in proportion to their financial contribution to societal wealth. However, a purely materialistic allocation not only privileges the privileged, but devalues non-material forms of societal contribution. Effort volunteered, for instance, to assist disabled and elderly members of the community, to assist single and employed mothers, to provide recreational opportunities for the urban young and to contribute to the quality of life of individuals and communities by such acts as planting and tending gardens may contribute to health and well-being, but not easily be quantified in monetary terms. If contributions to society are assessed only according to a dominant culture, contributions to promote the values of subordinate or minority cultures may go unrecognized and uncounted. Resource allocation to the value system of an urban society's political or governmental leadership may reflect inequitable biases and selectivity in what counts as a contribution that creates an entitlement to recognition and reward.

Outka says of his third principle of resource allocation, to each according to his contribution in satisfying whatever is freely desired by others in the open marketplace of supply and demand, that

> Here we have a test which, though similar to the preceding one, concentrates on what is desired de facto by certain segments of the community rather than the community as a whole, and on the relative scarcity of the service rendered. It is tantamount

> to the canon of supply and demand as espoused by various laissez-faire theoreticians (Outka, 1976, p. 87).

This principle appears more accommodating of differences among subgroups of an urban population, and more respectful of their distinctive characteristics. Its limitation is that, like the concept of "contribution" in the previous principle, it is too easily located in a materialistic or economic context, that is "the open marketplace of supply and demand." Health care services, therapeutic drugs, pollution controls and similar protections of the environment are unlike other services and commodities among which consumers can make choices, and it is inequitable that their difference should be disregarded. Public and clinical health care services and therapeutic products are valuable to vulnerable and sick people, maintenance and restoration of whose health and perhaps whose very survival depend on them. Accordingly, they should not be treated like discretionary goods and services that potential consumers may reasonably decide to forgo and among which they can exercise free market choice. Health is not an "optional extra" to life, but part of its essential quality and a common precondition to its continuation. Thus, to treat health care services and therapeutic products like common services and items open to choice of purchase violates the principle of justice that different cases should not be treated equally, without regard to the difference. This principle of health resource allocation offends the inverse of the fifth principle, which is that similar treatment should be provided for similar cases, by failing to recognize the special nature of health care services and products, and so treating dissimilar cases as if they were the same.

The fourth of Outka's principles, to each group according to its needs, comes closest to establishing how health care resources may equitably be allocated. It permits similar needs to be served in similar ways, and different needs of urban communities, and also of suburban and rural populations, to be served on the basis of their own elements. Further, when "needs" are distinguished from "wants" (Doyal and Gough, 1991), they may be determined on an objective, disinterested basis. This is not to say that any one theory or process of determination of relative needs is self-evidently right. Philosophers and, for instance, health resource administrators may invoke a variety of models of justice or fairness. The work of John Rawls, and his 1972 book A Theory of Justice has been significant in generating debate and analysis concerning how resources may equitably be allocated. Once an approach is adopted, however, it may be applied to determine and meet health care needs in a manner that can be shown to satisfy the theory of equity, justice or fairness that is professed. The chosen theory remains open to review and criticism as a matter of philosophy, including political philosophy, but practitioners of the theory can defend themselves against the criticism that they are motivated purely by self-interest or irrational bias, or that they have been thoughtless or indifferent to matters of equity.

The principle that similar treatment be available for similar cases appeals to a sense of justice in terms of logic, but when applied to health care it is a necessary but not a sufficient condition of equitable distribution of resources. The limit of its logic is that the principle alone does not require that any services be provided at all. The inequity of affording care in one case but not in another similar case may be remedied, in logic, by withdrawing care from the case in which it has been given. Equity in health and health care requires leveling upwards, however, not downwards, compatibly with the principle of beneficence. That is, there is an equitable duty to promote health and health services, since health is a positive good in individuals and populations.

A further limit of the principle is that similarly inspired responses to ill-health in population groups need not be identical responses. An adequate response, for instance, to an infection may be to treat all affected members of the population so that they recover. If the infection is non-lethal and self-limiting among populations of normal resistance, however, an alternative response may be implementation of preventive programs that reduce recurrence of infection in the community. Particularly vulnerable classes of people may be treated, such as the elderly or the very young, but others may be left of the processes of nature, and resources be prioritized to preventive care. Treating two infected groups with similar concern and application of resources does not necessarily require that they be treated identically (Wallace and Wallace, 1993).

The fourth principle, addressing the duty to serve population groups' health care needs, and the fifth, addressing the similar provision of care (but not necessarily the provision of identical care) in similar cases, may be combined or interwoven to create an equitable perspective on issues of health and health care raised by urbanization.

VI. HEALTH CARE EQUITY

The human right to maximize health has two aspects. The first is that health not be inequitably impaired or obstructed by the involuntary imposition of external forces, such as agents of pollution or contamination, for instance of the air, water supply or foodstuffs. The second is that, where nature and individuals' own resources are not adequate to resist illness, populations be supplied with health care and related services and supplies to maintain them at an equitable level of well-being. The first aspect concerns what are sometimes described as "negative rights", meaning rights to be left alone, and not be subjected to limitations on pursuit of one's own health or other preferences or to harms imposed by others without request or invitation. The second concerns so-called "positive rights", meaning rights to be afforded means to achieve health or other goals that cannot be achieved without aid (Cook, 1995, pp. 992 *et seq.*).

Some health interests depend on both negative rights and positive rights. For instance, interests in reproductive self-determination depend on negative rights not to be exposed to chemical or other sources of infertility, such as from toxic work-places or residential environments. Reproductive self-determination may also depend, however, on the supply of chemical or mechanical means of contraception, such as drugs taken by pill or injection, or on the fitting of intrauterine devices, or on the enactment and enforcement of laws such as empower women to resist premature marriage and to possess economic and other alternatives to repeated pregnancy and child-rearing (Cook, 1995; World Health Organization).

Urbanization may promote reproductive health and self-determination particularly in cultures where health depends primarily on positive rights to care, and where medications are most acceptably delivered by injection. Health care services may more easily be delivered in urban than in rural settings, because patients in more densely populated districts can reach centrally-located clinics served by teams of staff possessing a wide range of necessary skills. In contrast, rural populations are more likely to have access only to practitioners who visit their areas infrequently, and alone. Further, although oral medications may be distributed with similar benefit to urban and rural residents through pills that maintain efficacy for some time and are adequately packaged, many populations reject oral medication and will accept treatment only by injection. Among these populations, urban residents with convenient access to clinics may enjoy advantages that rural residents lack. Regarding injectable contraceptive drugs, for instance, urban dwellers may have access to birth control clinics, and so be better able than women in rural locations to plan and space pregnancies. Women's and infants' health is prejudiced by pregnancies that come too soon, too frequently or too late in women's reproductive lives, or too closely spaced (Cook, 1995). By reference to their wishes and needs, urban populations may have access to a good quality of care when, inequitably, rural residents do not.

VII. HEALTH FACILITY EQUITY

Rural hospitals and clinics may render a quality of personal care, through practitioners familiar with local needs, that compares favorably to the more impersonal care rendered in large, busy city-center hospitals whose principal staff live, for instance, in more congenial suburban areas. Regarding equipment, supplies and expert personnel, however, urban hospitals frequently appear superior to rural hospitals, raising an obvious issue of urban-rural equity. Even within urban areas, however, different hospital standards are found, which raises issues of equity among urban hospitals themselves. These issues are aggravated in countries where publicly funded metropolitan or municipal hospitals co-exist with private hospitals funded by higher-income recipients of services or their health care insurers. The contrast of public deprivation with private affluence raises central issues of social equity.

University or medical school teaching hospitals raise separate issues of equity that involve not only how hospitals treat the populations that depend on them, but also the reciprocal concern of how populations treat teaching hospitals. Some teaching hospitals are community based, for instance through suburban and rural facilities that train students for relevant styles of practice, but many are located in heavily populated city-center areas. They serve diverse populations, and have specialized facilities to which suburban and other hospitals refer patients for services that local community hospitals cannot provide. Often lacking a community that they serve exclusively, teaching hospitals function as regional resources rather than facilities devoted to individual populations, although local residents may make particular use of convenient access to them. As centers of excellence, they attract more highly qualified practitioners, and undertake research into advanced techniques, often with sophisticated equipment. Their presence among other urban hospitals may present different qualities of staff, equipment and specialized facilities, but not necessarily inequity or discrimination. It has been seen that, provided that individuals enjoy access to basic or essential health care for their needs, it is not discriminatory or unfair that some receive more. The superior, specialized care available in teaching hospitals does not render the standard of care routinely delivered in non-teaching urban hospital inequitable.

Recipients of more skilled or specialized care in teaching hospitals, reinforced perhaps with superior equipment, reciprocate in a way not required of patients in non-teaching hospitals. Much of their care is given by medical and other students under supervision, rather than directly by fully trained personnel, and care given by leading, highly qualified practitioners will be attended by trainees and perhaps directed in part to their instruction. Thus, patients who are treated in their own right may also serve in part as a means to the end of medical education. They are patients, but also "teaching materials."

It is a courtesy, and sometimes a legal requirement, that, in advance of medical students performing supervised procedures on patients or attending practitioners who treat them, patients be asked whether this is acceptable to them. If it is not, perhaps because they do not feel ready for the additional attention and stress of a particular occasion, students should be excluded from involvement at that time. If patients were to refuse in general in advance of particular treatments, or consistently on each occasion, to have students involved in their care, however, they might be transferred to non-teaching hospitals. The often unspoken equitable pact is that, in return for excellence of care available in teaching hospitals, patients make themselves generally available for the training of students. The pact is voluntary, in that patients unwilling to accept it will still be entitled to the general standard of care delivered in other hospitals. Further, if through local demand or referrals, a teaching hospital or one of its departments becomes overcommitted to patients who require a particular form of care, they may be transferred elsewhere or their non-emergency care may be postponed. Students must be exposed to a variety of procedures, in accordance with their training syllabus, and not spend their limited time for training in a field or speciality concentrating on a narrow range of procedures or pathologies that relate to the peculiarities of local or contemporaneous demand.

It is equitable that urban patients who receive the benefits of superior care in teaching hospitals bear the burden of providing teaching opportunities to students. The fact that rural patients and urban patients of non-teaching hospitals also benefit from the availability of medical and other health professionals who have been properly trained in teaching hospitals may be questioned but also explained on grounds of equity. The same applies to medical research, which tends to be more concentrated in university-affiliated hospitals than elsewhere. Patients there are more liable than others to be invited to participate in research studies and programs. They are free in principle to accept or decline the invitation and the associated risks. In reality, however, they may feel pressure, and be subject to conscious or unconscious pressure or undue inducement, to join studies from the results of which patients who do not participate may benefit. The equity of relations between those who participate as subjects of teaching or research, who are particularly liable to be urban dwellers, and beneficiaries of teaching and research who do not participate, may be more effectively explored, and established, at the level of macro-ethical analysis than at the level of person-to-person ethics.

VIII. WIDER DIMENSIONS OF EQUITY IN HEALTH CARE

The modern growth of cities and megacities, not confined to any single region or hemisphere of the world or level of national industrialization (Saleem, 1995; see p. 9 for leading urban growths), reflects inequities between rural and urban life. Even in relatively affluent and well-served countries, rural health care facilities can be severely disadvantaged in the care they can offer populations that depend on them (see Kiel, 1993 on the situation in the United States). Individuals and families must determine whether the advantages of rural life that they enjoy are compensation for the disadvantages. People move to cities for such reasons as to improve their economic, educational, employment, housing, health care and related opportunities. It is a matter of policy for each country or region to determine whether to resist movement of rural populations to cities by improvement of the facilities and quality of rural life, narrowing inequities or enhancing the compensating advantages of rural life, or to meet the expectations of urban residents by provision of the housing, sanitation, health care and other services that attract people to cities.

The status of suburbs to major cities is ambivalent, and may come to warrant significant attention regarding equity in health and health care. Suburbs may be or become relatively neglected urban areas for provision of services, for instance, if they are only "dormitory suburbs" where populations go at night, traveling long distances to remote, underserviced areas where residential land is cheaper, but depending for employment, shopping, higher education and health care on resources at city centers. In contrast, however, suburbs may compete successfully against decaying city centers, where poverty and crime rates are high among abandoned residential structures, shops and factories. The flight of population to more congenial suburbs, reflecting and aggravating the "hollowing out" of city cores, may at first expose the scarcity of public health care resources in suburbs, but later reverse the imbalance by causing health care facilities also to abandon city centers and relocate among suburban populations (Wallace and Wallace, 1993).

Urbanization has further implications for national and international equity in health and health care. The spread of disease among urban populations, including suburban populations, in conditions of widespread regional, national and international travel and immigration risks the spread of new and exotic diseases to cities unfamiliar with their nature, and ill-equipped to respond to them with appropriate personnel, drugs and facilities. The AIDS pandemic, for instance, first appeared as an urban phenomenon, and in some areas a virulent strain of tuberculosis, resistant to penicillin, is following that pattern of spread. Cities in temperate regions, for instance, may house residents infected with tropical diseases, through immigration and tourism.

Further, rapid urban growth can lead to the paving of agricultural land and cutting of rain forests to provide human habitations, with consequences for the environment and ecology of regions far beyond the limits of expanding cities that affect health. One response to dependency on foodstuffs grown at more remote distances from where urban populations live, brought in at great cost and difficult to preserve in a fresh condition, is urban agriculture. It has been noted that:

> From the garbage dumps of Calcutta to the rooftops of Mexico City and dirt strips along roads in Jakarta and Nairobi, millions of people are farming the world's cities, feeding their families and supplying markets with vegetables, fruit, fish and even meat . . . Urban agriculture is a rapidly growing industry, providing a significant amount of the world's food . . . The ability to farm, however small the plot, can contribute significantly to family health (Crossette, 1996).

This resourceful way to maintain health in urban populations may increase rural movements to cities, further impoverish rural life and reduce exportation of commercial foodstuffs, for instance, from less industrialized to more industrialized countries and regions. These effects will trigger a further cycle of concerns about urbanization and equity in health and health care.

Faculty of Law,
Faculty of Medicine and
Joint Centre for Bioethics
University of Toronto
Toronto, Canada

BIBLIOGRAPHY

_______: 1996, 'Managing Asia's Cities', *Environmental Policy and Law* 26(1), 23.

Cook, R. J.: 1994, *Women's Health and Human Rights*, World Health Organization, Geneva.

Cook, R. J.: 1995, 'Human Rights and Reproductive Self-Determination', *American University Law Review* 44, 975–1016.

Cross, F. B.: 1995, 'When Environmental Regulations Kill: The Role of Health/Health Analysis', *Ecology Law Quarterly* 22, 789–784.

Crossette, B.: 1996, 'More People Grow Food in Cities, Report Says', *The New York Times*, February 18, reporting on *Urban Agriculture: Food, Jobs and Sustainable Cities*, published by the U.N. Development Program for the Habitat II conference held in June 1996 in Istanbul, Turkey.

Doyal, L. and Gough, I.: 1991, *A Theory of Human Need*, Macmillan, London.

Goldstein, G. and Kickbusch: 1996, 'A Healthy City is a Better City', *World Health* (publication of the World Health Organization).

Gostin, L.: 1991, 'Macro-Ethical Principles for the Conduct of Research on Human Subjects: Population-Based Research and Ethics', in Z. Bankowski, J. H. Bryant, and J. M. Last (eds.), *Ethics and Epidemiology: International Guidelines*, Council for International Organizations of Medical Sciences, Geneva, pp. 29–46.

Kiel, J. M.: 1993, 'How State Policy Affects Rural Hospital Consortia: The Rural Health Care Delivery System', *Milbank Quarterly* 71, 625–643.

Mahler, H.: 1987, 'The Safe Motherhood Initiative: A Call to Action', *Lancet*, 668.

National Centre for Health Statistics: 1988, *Advanced Report of Final Mortality Statistics*, Monthly Vital Statistics Reports, September 30.

Outka, G.: 1976, 'Social Justice and Equal Access to Health Care', in R. M. Veatch and R. Branson (eds.), *Ethics and Health Policy*, Ballinger, Cambridge, Massachusetts, pp. 79–98.

Powers, M.: 1996, 'Forget about Equality', *Kennedy Institute of Ethics Journal* 6, 129–144.

Rawls, J.: 1972, *A Theory of Justice*, Oxford University Press, Oxford.

Rees, W.: 1996, 'Ecological Footprints of the Future', *People and the Planet* 5(2), 6–9.
Saleem, O.: 1995, 'Be Fruitful, and Multiply, and Replenish the Earth, and Subdue It: Third World Population Growth and the Environment', *Georgetown International Environmental Law Review* 8, 1–43.
Wallace, R. and Wallace, D.: 1993, 'The Coming Crisis of Public Health in the Suburbs', *Milbank Quarterly* 71, 543–564.
Wigley, D. C. and Shrader-Frechette, K. S.: 1996, 'Environmental Racism and Biased Methods of Risk Assessment', *Risk: Health, Safety and Environment* 7, 55–88.
World Health Organization: 1994, *Women's Health and Human Rights*, Geneva.

THOMAS J. BOLE, III

ENGELHARDT ON KANT'S MORAL FOUNDATIONS AND HEGEL'S CATEGORY OF THE STATE

I. PREFATORY

Stuart Spicker's genius, as Laurence McCullough has remarked, is that in a half-hour conversation he can point to at least a half-dozen implications in one's ideas, each worth pursuing, and none of which their owner had previously realized. "Critical thinkers are commonplace," McCullough epitomized, "original ones are not."

Scholarly fields exist, Ludwig Fleck observed, only if they have a fabric of discourse to sustain them. Over the last quarter-century Stuart Spicker has helped create and sustain the discourse of bioethics, co-editing the more than 50-volume *Philosophy and Medicine* series, and now the *HEC Forum*. That there is in the field a balance of logico-lingusitic-analytic, broadly positivistic, with non-analytic concerns that attend to the phenomenological contexts that shape meaning, is largely due to Stuart Spicker's influence. His stress on the importance of philosophical anthropology has pointed to the value-laden contexts within which scientific medicine labors (ten Have, 1997). He has thus drawn our attention to the task of philosophy in medicine, to explicate, explain, and evaluate these values (Spicker, 1990). These efforts have also called attention to the subtler phenomenological contexts which bioethical analysis methodologically disregards, but which are necessary to uncover analyses' tacit presuppositions.

Spicker also recognizes of the fact that philosophical anthropology is not the only conceptual context necessary to understand philosophy of medicine and bioethics. Philosophical anthropology does not attend to the human qua "personal agent" (Spicker, 1975, p. 198). This is the aspect of man most relevant to the theoretical framework of bioethics, and – insofar as human agents are intrinsically social – to the social and political philosophy necessary to understand health care policy. In what follows I should like to analyze critically the bioethical and social- and political-philosophical thought of Spicker's co-editor in the *Philosophy and Medicine* series, H. T. Engelhardt, Jr., as articulated in

H. T. Engelhardt, Jr. (ed.), The Philosophy of Medicine, 195–221.

the second edition of his *Foundations of Bioethics*. Because no one has had as many philosophical conversations with Spicker as his closest friend Engelhardt, I trust that Spicker will enjoy, and hope that he will not have already foreseen and rebutted, some of the implications of Engelhardt's philosophy I try to draw.

II. THE CRITICAL BASIS OF ENGELHARDT'S FOUNDATIONS

Engelhardt critically questions the secular moral warrant for public policy, political or legal, unless it is licensed by the actual or clearly implied consent of those upon whom it is binding, or is directed against those who use force against unconsenting innocents, i.e., force that is not so licensed. He does not think that theory can provide a basis in reason, and so normative for the relevant moral agents, for enforcing morals or public policy. Sound rational theory cannot legitimate either one notion of morals rather than some other or one notion of social or political arrangement rather than some other – unless it depends upon premises with which one may reasonably disagree. Nonetheless, where one's reason cannot do so, one's will can, by deciding to live peaceably with others and therefore not to engage in actions using unconsenting innocent others without their consent. This is the principle of permission. It provides a procedural basis binding all moral agents to the ideal of peaceable co-existence. The principle for content in morality is that of beneficence. This gives content to morally binding goals of action. It is morally authoritative, however, not because it is cogently argued for, but because it is responsibly consented to.

The theoretical center of his book thus has two parts, one negatively critical, and the other positive. The critical center is the argument that secular reason, i.e., philosophical theory, cannot show the normativity of any moral content. It is a critique that is also intended to undermine any claim that states have secular moral authority to enforce positive policy that does not rest upon the explicit – or clearly implicit – consent of those moral agents whom the states govern. The radicality of the criticism has not in general been taken seriously by the philosophical community, as the introduction to the second edition of his *Foundations* makes clear. I want to try to do so now.

Engelhardt's contention is that one cannot in principle resolve moral controversies by sound argument, because one cannot rationally defend the (set of) first premise(s) – the ultimate foundation in virtue of which a particular content is claimed to be paradigmatically moral – of one's moral theory. There is no way to argue in behalf of the ultimate standard determining moral content except by appealing to a higher standard of moral content. Any standard "presupposes what it would justify: a particular moral content" (Engelhardt, 1996, p. 42), about which moral strangers can reasonably disagree.[1]

Engelhardt concludes that there is no way to justify completely the content-filled first premises of moral theory. He infers that there is no way to justify in terms of reason alone the imposition of any particular moral view or public policy upon unconsenting innocents in a secular pluralist society. The only way to justify actions that involve innocent others who may be moral strangers is the contentless, purely procedural one of making this involvement dependent upon their permission.

The two features of Engelhardt's position to which I shall attend may be called the Neokantian and the Neohegelian strands. They reflect respectively Kant's transcendental argument for the foundation of morality involving others in a secular pluralist society, and Hegel's notion of the state as that structure within which individuals, communities, and economic and social classes, which may be moral strangers to each other, can live an ethical life. Engelhardt must also appeal to a transcendental argument to ground his ethics of actions involving innocent others who may be moral strangers. Moreover, he makes this ethics the basis for a reconstruction of the Hegelian state that is "libertarian by default." In the rest of this section I explicate each strand, in preparation for evaluating them in the next two sections.

The Neokantian strand is Engelhardt's self-styled transcendental argument (Engelhardt, 1996, p. 70) to ground peaceable intercourse with other moral agents, even though they may be strangers to one's own basic moral premises. One cannot ground first principles in prior principles; so one argues self-reflexively, in terms that one's disputant about first principles must presuppose in order to engage in the dispute. Classical examples of non-transcendental self-reflexive arguments are Aristotle's apagogical argument for the principle of contradiction (in *Metaphysics* IV.3–4) and Descartes' argument about the indisputability of the *cogito ergo sum* (in his second *Meditation*). Though they cannot be disputed coherently without affirming them, they are not transcendental because they are, respectively, simply objective and simply subjective. The peculiarity of transcendental arguments is that they claim to argue in terms of presuppositions not for objective, nor for simply subjective, states of affairs, but for the subject's normative performances, theoretical or practical, upon objects. A transcendental argument tries to show why some (set of) conceptualizable structure(s) is necessary for knowledge claims to be objectively true or, as in our case, for acts to be morally good, or in accord with what is morally obligatory.

For Engelhardt the transcendental argument is necessary to understand how in a secular pluralist society one can act morally in ways intended to affect others to which they cannot be assumed to assent. We do act in ways that involve others, and it is practically necessary that we do so, unless we are hermits. "If one is interested in collaborating with moral authority in the face of moral disagreements without recourse to force, then one must accept agreement among members of the controversy of peaceable negotiation as the means for resolving concrete moral controversies" (Engelhardt, 1996, p. 68). The *apodosis* follows, because one must act in ways that involve others, and in a secular society one is liable to do so with others who are moral strangers. This means that one's action involves using force against them and needs justification. Reason cannot provide it. The will to live morally, however, can: if one wants one's actions involving innocent moral strangers to be morally authorized, their permission is the necessary – and, given the failure of ethical theory, sufficient – condition for doing so.[2] The transcendental argument in fine is that, in order to justify actions involving innocent moral strangers, one must, given the failure of reason, will not to involve them without their permission.

Notable is that the argument straddles the traditional domain of ethics, which is concerned with values (e.g., in Aristotle's *Nicomachean Ethics*) or norms (e.g., Kant's doctrine of virtues) which any moral agent should act to achieve, whether in his own life or with respect to anyone else, and social ethics and social and political philosophy (e.g., Aristotle's *Politics*, Kant's doctrine of law). The latter is concerned with justifying those normative social concepts, and most notably the concepts of the state and its laws, which insure the actual coexistence of the freedom of different individuals in society, The former is socially underdetermined, because it does not take account of social institutions – notably, the state and its laws – and their normativity.[3] To take the example of Kant, whose transcendental argument will concern us immediately below, an action is moral if it is willed in accordance with a maxim which can at the same time obtain as a principle establishing universal law for the will of any moral agent. The universalization is oriented to what human persons will rationally in order to actualize their self worth. It is assumed that the universe of all maxims conform, that each moral agent can see what is rational for all, and that there will be no problems in conflicting determinations or points of view. The kingdom of ends is simply the sum of conforming ends abstracted from any concrete societal structures. By contrast, the law as a social norm is more determinate than ethics because it must take account of social pluralities (as well as person-to-person cases such as contract); but it is also less determinate because it is concerned with the minimal norm which must be followed and enforced, and abstracts from whether or not motivation for following the norm is ethical or prudential. To put the contrast briefly if crudely, ethics is addressed to the ground of responsible action by any moral agent, his conscience or free will, and concerns the motivation for every man's actions, whereas social philosophy and social ethics is concerned with the concrete execution of individuals in social settings, and with the individual and institutional norms necessary to ensure that individuals can coexist in freedom of action. Engelhardt makes his ethics address not, as with Kant, the ground of moral obligation, the moral agent's free will or conscience, but actions in a social sphere where it may involve moral strangers. And given the incapacity of reason to establish morally canonical content, and therefore content for public policy, he makes ethical individuals totally explanatory of social institutions.

The second noteworthy feature is Engelhardt's Neohegelian interpretation of the state. The state is, as for Hegel, that unity which provides the actual, and self-sufficient, framework within which communities, and individuals, with diverse moral understandings can live together peaceably, within the framework of laws (cf. Hegel, 1967, §§ 258, 260–261). And yet it is libertarian by default. It cannot serve the common good of liberty, for this good, like every good, cannot be shown normative for moral strangers in a secular pluralist society. It has no legitimate purpose other than to provide the procedural protections and welfare rights to which the diverse communities and individuals have either explicitly or tacitly consented (Engelhardt, 1994, p. 218). To understand why, we need to look more closely at Engelhardt's treatment of the legitimate authority of the state and, as a necessary prerequisite, his treatment of property.

"The principle of ownership (as well as the principle of political authority) is a special expression of the principle of permission" (Engelhardt, 1996, p. 164). Ownership reflects the senses in which persons articulate their free will in and through their bodies and in things – and animals – which they produce. "[I]nsofar as persons (1) extend themselves to objects by making those objects theirs by transforming them into products, (2) acquire rights to the person or body of others through the consent of those persons, or (3) have transferred to them rights in objects or persons, they have property rights that must be respected as a part of general secular morality" (165). Communities and societies only have property given them by the consent of the persons constituting them.[4]

Engelhardt wants to argue that for secular pluralist societies, states only possess authority and resources to which those who fashion the states have consented (Engelhardt, 1996, pp. 167–172). One cannot base government authority upon a common good grounded upon the goal of achieving the "morally canonical concrete understanding of the good life and moral obligations" (Engelhardt, 1996, p. 168), because of the earlier arguments against theoretically grounding such a canon. Nor, for the same reason, can one appeal to a hypothetical contract or to hypothetical contractors, as Rawls does. Moreover, an actual contract does not obligate those who are not signatories, e.g., those in subsequent generations, for these have not consented. Nor is prudence a possible justification of governmental authority (Engelhardt, 1996, pp. 169–171): we submit to government coercion because the alternative is even more coercion from marauding armed bands.[5] Coercion is not consent; Alexander the Great's power is not more legitimate than the outlaw band's because he has greater firepower. The only source for the moral authority of government is "the actual consent" of the governed to the government's actions. "Unconsented force can be forbidden, contracts enforced, and health care provided as a welfare right out of commonly held resources. [But] such state secular moral authority will not extend to the control of the consensual action of free individuals, including their use of their private property" (Engelhardt, 1996, p. 171).

Some issues that cannot be resolved by the agreement of all because there are neither sufficient moral premises held in common nor sufficient common understanding, are issues such as establishing property boundaries, determining punishments for crimes, and specifying courts' jurisdictions. About these one must "do the best one can and resolve controversies by means that involve the least coercion and the greatest amount of consent". Market mechanisms are best adapted to such means, Engelhardt thinks, because they result from numerous acts of consent by individuals who do not share a common moral vision. "Where such are not sufficient, robustly limited democratic mechanisms, limited by eschewing any goal but the stipulation of points of unclarity, will need to be employed" (Engelhardt, 1996, pp. 171–172).

The crucial point is that the state cannot in principle serve a common good that is practically superior to the goods of the individuals or particular parts of it, because no such good can be morally warranted. The moral state is as amoral as the post office, which serves equally the postal needs of various individuals and communities without representing or censoring the views of any. More precisely, the moral state should be so amoral that its postal service would only exist if actually consented to by the citizenry, not as a legally protected (in some respects, at least) monopoly. The only difference between the morally authorized state for Engelhardt and the morally authorized corporation existing to pursue common goals agreed upon by its members is not that the state alone can back up its laws by force – corporations, too, could licitly use force upon its members pursuant to what the latter contract – but that the state encompasses different economic corporations, different moral communities, and even moral strangers consenting to live peaceably with each other according to the rule of law the state provides. In this sense the state is a second-order corporation, a morally authorized social force within which individuals and groups with diverse ideas of the good, non-economic as well as economic, can protect their peaceful co-existence.

I find objection with both the Neokantian and the Neohegelian strands. The first is Engelhardt's claim that successful transcendental argument in morality yields valid procedure, but not valid content, contrary to what Kant thinks. I shall contend that Kant's argument is the more fundamental to the concept of morality, and that if we distinguish the categorical imperative in its function as defining the moral law from its function as a criterion for the morality of its maxims, we have no problem seeing how it is content-filled. Engelhardt's Neohegelian point is that the morally authorized state is libertarian by default. I shall contend that in delegitimizing the common good, Engelhardt denies authorization of public functions necessary to prevent the use of the state to support the particular interests of the strongest of those particular persons and particular communities that constitute it. On this reading his political theory has authoritarian consequences inimical to libertarianism.

III. DEFENDING KANT'S GROUNDING OF MORALITY

Engelhardt consciously follows Kant in his appeal to transcendental argument in order to explain what is supposed to be an unavoidable feature of our moral lives. A comparison of the features which he and Kant are each endeavoring to explain will suggest that Engelhardt is incorrect to claim that one cannot morally ground content. The comparison is necessary to evaluate effectively Engelhardt's criticism of Kant's attempts to validate determinate moral content. It is also necessary to set the stage for the next section.

Though Engelhardt criticizes moral theory, and in consequence the moral authorization of the coercion consequent upon public policy, in general, it is well to remind ourselves of his dominant concern in the *Foundations*. It is to ground moral judgment in bioethics "in principles for resolving moral disputes among individuals who do not share a common moral vision" (Engelhardt, 1996, pp. 121–122). This is the feature his transcendental argument is designed to explain. Moral theory, or reason, cannot provide the principles sought, but the will to live peaceably with others can. If one wants to live in peaceable community with innocent others who are used in one's actions, one must not use them without their permission.

Engelhardt's approach to bioethics is striking in that it is an approach to ethics and social policy in general (cf. Engelhardt, 1996, pp. 35–84), and as such it abstracts from the most basic datum of morality, viz., that we should be – are categorically obligated to be – moral.[6] It is this datum that is the focus of Kant's concern (cf. Kant, 1964). We are morally responsible because we are practically rational. As practically rational, we even bring the willing of individual objects (e.g., the eating of this apple before me) under a rule for action, what Kant calls a maxim (e.g., I shall eat this apple because I like the taste of this sort of apple). Most of our maxims are maxims for the pragmatic, or instrumental, goodness of our actions. The morally good, however, is not instrumental, but absolute. Just as we are interested in the set of general criteria for the truth of our knowledge claims, so we are interested in the set of general criteria that make our willing and acting morally good. These constitute the basic principle of morality, the moral law. This is a primary example of a standard which Engelhardt thinks rationally indefensible. It is also the basic principle which Kant is concerned to articulate and defend (Kant, 1956; 1964).

The source of the moral law, and of morality's categorical imperativity, is practical reason, not in its empirical uses, but "pure", concerned to achieve not what is good functionally but what is good absolutely. Pure practical reason is the good will, the moral will. It dictates the categorically imperative moral law, "Act only on that maxim which you can at the same time will that it should become a universal law" (Kant, 1964, p. 88 [AK IV, p. 421]; cf. Kant, 1956, p. 30 [AK V, p. 30]). The moral law is realized in any moral agent orienting its willing and acting to the demands of morality, because the moral agent thus determines itself as autonomous.

Aschenberg makes a valuable distinction between the function of the categorical imperative to define the moral law, or what Kant calls the "fundamental law of pure practical reason" (Kant, 1956, p. 30 [AK V, p. 30]), and its function as a criterion by which to judge which of our rules for action, our maxims, are moral (Aschenberg, 1996, p. 29).[7] Its criterial function is that with which we must compare Engelhardt. First, though, consider the defining function. It is the elucidation of why we should bother with morality at all. As Engelhardt remarks, the sanctions against immorality are intellectual; "[t]o raise an ethical question is to raise an intellectual question regarding justifications for action" (Engelhardt, 1996, p. 108).

One notes that morality *strictu sensu* deals primarily with the motivation of one's willings and actions – what Kant calls the form – in contrast to their goals, results and consequences. These are dependent, at least in part, upon the contingencies of the situation within which one wills and acts; the former is the agent's total responsibility. Morality consists of willing (and acting to do) good because it is good. One must will not simply in accord with duty but out of duty. I might help the crippled beggar out of sympathy, or of peer pressure, or for the sake of my social reputation. All of these are amoral motives, and in other situations can move us to act immorally. I act morally if I act out of the consciousness that as a rational agent I ought to do so; the act helps the cripple meet basic needs so that he might better function as an autonomous being, and it promotes a social practice which, prudently pursued, encourages such help to its members. This obligation receives content from the *humanitas* formula, to respect other rational agents as "ends in themselves," and not to be used as mere means; about which more below, as well as about the analogous example of meritorious duty in Kant's *Groundwork*.)

An observer, of course, can never know whether an action which is to all appearances morally right, is grounded solely on a moral maxim, or also on some other. Kant observes that one cannot be sure that one's own actions are not entirely morally motivated, with no tincture of self-love (Kant, 1964, p. 74 [AK IV, p. 407]). (That we can so motivate our actions is presupposed by the fact that we stand under the imperative of our practical reason to do so.) But how is the Kantian framework relevant in bioethical contexts of grounding judgments to moral strangers, where motive is secondary? The Kantian answer is that it is explains why we are free, morally autonomous: because we stand under the imperative to act morally, an imperative which we issue to ourselves via our practical reason.

This answer gives a justification for morality, and a sanction for immorality, which is not available to Engelhardt. Engelhardt explains his sanction for immorality as follows:

> To use unconsented-to force against the innocent is incompatible with holding that others are wrong in using such force against oneself, or meaning anything more by terming another wrong or blameworthy than that one dislikes the other's conduct and wishes that he and others would refrain from it. In short, the notion of a peaceable community as fashioned by the principle of permission is a cardinal element in the lives of persons (Engelhardt, 1996, p. 108).

But this does not argue against one whose fervent embrace of freedom seeks to be free even from the constraints of respecting an innocent other who might be caught up in his actions and who asks why they should be worthy of respect. He is willing to be consistent, and to hold an intersubjective standard of right and wrong: right is the realization of the free man's designs. He simply calls into question Engelhardt's notion of a peaceable community insofar as it conflicts with those designs. For this amoralist there is no reason to curb the prudent pursuit of what he wants. (It is not an objection to say that he cannot give cogent argument for his case to others; adequate answer is that they have no cogent argument against him realizing his freedom.)

Kant, however, does have an argument: he can point out that such an idea of freedom is negative; it is defined by that from which it would be free. Positive freedom, however, is acting rationally, a *sine qua non* of which is acting according to moral maxims. This is the reason why one should be moral, and why one should respect others, even if one does not assume as a basic premise that the peaceable moral community is good.

It remains to be shown, however, that Kant's categorical imperative can effectively act as a criterion. The criterial use is of the categorical imperative is not a use of a major premise. Rather, it is the use of the form by which to judge which among the countless goals of our will can be willed out of respect for duty. These goals are countless because they are constituted by the material or object willed, which is itself dependent upon the factual contingencies of the demands of our factual makeup and social surroundings. The contingencies cannot help constitute a normative, universally valid criterion of morality; rather, given the categorical imperativity of morality, they presuppose such a criterion in order to understand how that imperativity operates in them.

Consider two of Kant's criterial formulations of the categorical imperative: that you should act as if the maxim of your action could by your will be a universal law governing the nature of human relations (Kant, 1964, p. 88 [AK IV, p. 421]; the "law-of-nature" formulation); and you should treat humanity, whether in the instance of your own or another's person, as an end in itself, and not as a means merely (Kant, 1964, p. 96 [AK IV, p. 429]; the "end-in-itself" formulation). The latter first.

Like Engelhardt's principle of permission, it forbids any maxim, i.e., considers it immoral, that treats any person as a means merely. For Kant, however, this formulation is issued by reason to any rational agent, including the actor: It is immoral to treat myself as a means merely, e.g., habitually to satisfy an addiction for whiskey or drugs rather than to develop my intellectual talents.[8] It is immoral in my own case, because universalization of the maxim by which I would treat myself as a means merely, even with my consent, eschews my responsibility for the action. Engelhardt would deny in principle that content-filled ends can be universally defended such that the agent, whether oneself or another, could not in principle set them aside. Consequently, he would also deny there are secularly enforceable public duties to treat oneself as an end, or even positively to encourage such autonomy and/or to discourage its self-destruction.

Set aside the consequence for the next section, in which conditions of realizing a peaceable secular pluralist society are discussed, along with whether the societal structures have the right to encourage some forms of activities and discourage others, e.g., the victimless crimes of drug addiction. Even at the non-institutional level of interpersonal relationships, it seems evident that we are able in principle to distinguish the sorts of ends that are compatible or encouraging of a moral subject's responsible activity from those that are not. Otherwise we could not distinguish between respecting one's actions to destroy the factual conditions of his being a subject by indulging excessively in the consumption of drugs, and respecting his moral subjectivity by discouraging that abuse. Our endorsement of the end-in-itself criterion does not imply that we should endorse a blanket ban on suicide, as Kant suggests (AK IV, 429). Kant allows that the morally obligatory cannot always be legally obligatory (Kant, 1991). The criterion is necessary, however, to distinguish between cases of the individual about to jump off the bridge because he has just been diagnosed terminally ill and had his lover leave him, and those in which, like Sigmund Freud in the terminal stages of cancer, one is asking for a fatal amount of morphine because the chronic pain makes further rational activity impossible. No matter how we judge the latter sort of case, we could not in any way distinguish it from the former without the end-in-itself criterion.

More interesting, and certainly more complicated, is the natural law formulation, "[a]ct only on that maxim through which you can at the same time will that it should become a universal law" (AK IV, 421). A subjective maxim is valid, can be a part of the content of objective moral law, if and only if it can be conceived without contradiction and rationally willed[9] to be an inviolable law for the interactions of human persons.

On the one hand, universality is an essential mark of the rationality of the maxim; so this formulation is indispensable. In this regard, it functions essentially differently than generalizing a situation in terms of one's desires and aversions (cf. Wagner, 1992, p. 378). On the other hand, Kant is too optimistic to think that this criterion can be always applied without empirical specifications in order to test whether one's practical maxims conform to the moral law. One may argue that the strict duty not to lie to others does not apply in the case in which the other has no rightful claim to the knowledge involved in telling the truth, and would, were he exemplified in the person of a pursuer bent on murder, effectively accomplish some evil with that knowledge (cf. Aschenberg, 1996, p. 37). The strict duty not to lie might admit of exceptions under certain conditions without being undermined as a moral law for the interactions of persons. At any rate, it is difficult to see how the natural-law formula alone suffices as a criterion for differentiating moral from immoral maxims. For, apart from idealized situations, the criterion would seem to need the practical skill of correct application to the situation at hand. In addition, there is the question of the relevance of the probable consequences of the action motivated by a maxim meeting the criterion of the natural-law formula: though the consequences do not touch the morality of the motivation, they may bear on the intelligent application of the criterion, as in the example of the pursuing killer. Despite these problems, the natural-law formulation is a necessary reflection that the categorical imperative, and the moral law which it expresses, is normative for all moral agents.[10]

IV. DEFENDING HEGEL'S CONCEPT OF THE STATE AGAINST ENGELHARDT'S LIBERTARIANISM BY DEFAULT

Engelhardt's contention for a minimal moral order depends not only upon the destructive critique of the soundness of arguments – in terms cogent in a secular pluralist society – to establish a contentful good. It also depends upon those structures really necessary for individuals and communities who may in fact be moral strangers to each other to live in peaceful co-existence. Such structures are most obviously instanced in free markets and more controversially in states, social formations that claim the right to a monopoly on power in order to implement effectively just laws.

A notable upshot of Engelhardt's destructive critique, however, is that states are said to possess no moral authority not possessed by individuals: "states may do only what any individual may do to secure the mutual respect of persons, to support the discharge of obligations of beneficence, and to protect the property rights of individuals and communities" (Engelhardt, 1996, p. 167). On this view the state is not justified as the necessary support for the realization of the good life in the world of practice, as Aristotle thought. There is no content-filled common good distinct from that to which the individuals governed each explicitly or implicitly consent.[11]

I want to argue that such a picture of the state does not secure a framework within which individuals, or communities of individuals sharing the same or similar view(s) of the good, can really live in peace with moral strangers. For the various individuals and communities share no common goal; they legitimate state functions not in virtue of being citizens, but rather, *vis-à-vis* each other, in virtue of being self-interested individuals. Each party would, out of self-interest, legitimate state functions only if these serve his (or his community's) self-interest. State power in principle should serve no interests save those of preserving rights of particular individuals or communal parties simply as ultimately self-interested, and taking care of coordination problems. It should not serve the interests of those parties considered as *ab initio* committed as citizens to some overarching common good. Then the moral imperative not to use state powers to pursue one's own interests against the consent of others is the lone theoretical underpinning authorizing the state's claim to a monopoly on power. It would follow that either this power is used with moral restraint, in which case there is no need for the state, or it is used selfishly or partisanly, i.e., despotically, which seems more likely. I shall contend that Engelhardt's libertarianism by default licenses the Moloch state.

This contention is established by showing the defects in Engelhardt's adaptation of Hegel's concept of the state in the *Philosophy of Right* to Engelhardt's libertarianism by default.

A. *The Project of Hegel's Philosophy of Right*

Hegel's *Philosophy of Right* treats the domain of philosophy he aptly calls mind (*Geist*) objective in the outline of his entire system of philosophy in the *Encyclopedia of Philosophical Sciences*. It is the articulation of those normative structures of man's practical life: individual property and rights, morality, and societal structures such as family, economic and civil society, and the state. Its overarching structure is the articulation of the will as both really free and rational. Because the will is construed as practical reason, it is a universal principle, one by which those who have it – mature humans – can understand how, in virtue of having reason, they can assent to goals, instanced as concrete social structures, most notably in law, which are supra-individual.

To understand the will as free, Hegel wants to say, one must first understand how the individual can realize his will in things about him: by taking them into possession and using them to express his will, and thus exercising his claim of right to them; by mutual recognition of these rights in exchange, the common value of which is expressed in contract for the parties involved; and by clashes between what is right in itself, the common value contracted, and the particular will of a contracting party, which contradicts that right and thus instances the concept of wrong. The stages of wrong are graded from non-malicious, or civil, to criminal and coercive. In the former the disputants recognize a common right implicitly, but disagree on its expression. In crime the particular criminal recognizes, in denying, a universal right and objective expression of my will. Retribution is exacted by the particular party wronged, as revenge, it is not going to express the notion of what is universally right, but rather a new wrong. The transition to the level of morality is to the level at which right is universal because explicitly reflective, or rational, even though expressed by the particular subject (cf. Hegel, 1967, #103). The individual person recognizes that he is explicitly free in the reflectively rational sphere of the subjective, of the moral, to which he should subordinate all particular manifestations of will. To understand the will as free, Hegel is saying, it must also be understood at a second and more complicated stage as universal, because rational, yet subjective. This second stage contains the treatment of Kantian morality.

Hegel develops the concept of morality so as to explicate the difference between form and content. The form is that which in my purpose in willing, for which I am wholly responsible, and if moral it is concerned to will what is right. What I purpose, the content willed, is actualized or executed in particular circumstances. In it one must distinguish the universal purpose intended, to do what is right, from the particular goal, to realize what is subjectively good in the particular case (cf. Hegel, 1967, #114), the subjective welfare. Universal intent and the subjective welfare are united in the notion of the good, at that which every will aims *qua* rational, and which is good for all rational wills, and as that which addresses me insofar as I am reflectively free, i.e., in my conscience.

This is Kant's moral law, "the Idea as the unity of the concept of the will with the particular will," i.e., as the good will in its function as categorically imperative upon the particular will. "The good is thus freedom realized" (#129). The good of the moral law makes it one's duty in particular cases to realize what is good not simply for oneself, but for all relevant others, to "strive after welfare, one's own welfare, and welfare in universal terms, the welfare of others" (#134). But because this good – the law for any moral agent – is not realized, Hegel develops the disparity between it and moral subjectivity, or conscience. Here Hegel praises Kant for having first recognized that duty, or moral obligation, and therefore the moral law or moral good that is its objective correlate, are constituted by the particular moral agent or "subject" (#133), not *qua* particular, but *qua* recognizing his will as rational. As rational

I can abstract from my, or others', particular interests and/or claims of right or wrong, and ask, "Is it right, or good, not in terms of this or that subject, but simply in terms of what it is rational to do or to enact?" He goes on, however, to condemn Kant, because the standpoint of the universal good is formally disconnected both from the fulfillment of my will, and from particular societies within which right and welfare are specified. The moral subject can never be sure that his rational projection of the good is only private. (Hegel can then explain both evil, as the conscious choice of my good over what I ought to do, whether done in bad conscience or in a hypocritical one (##139–140).) The good, as what free wills should actualize as rational, and the subjective freedom of conscience, mutually implicate one another. But one cannot understand this mutual implication in terms of the individual moral subject, or in terms of the concepts of morality. Rather, this mutual implication can only be understood in terms of the concepts of "ethical life" [*Sittlichkeit*], i.e., in terms of concepts that articulate those social formations within which the individual actualizes both his moral duties and his subjective freedom.[12] If free will – freedom – cannot be completely understood unless it is understood as both rational and objectively realized, then it cannot be fully understood either in terms of individuals expressing their freedom (concepts of abstract right) or of freedom as rational, conscientiously reflective, or simply as subjective (concepts of morality), but in terms of concepts that articulate those social formations within which the individual actualizes both his moral duties and his subjective freedom, the concepts of *Sittlichkeit*.

The concepts of *Sittlichkeit* articulate supra-individual structures which are unified by being actualizations of practical reason. They are normative concepts, because they are unities encompassing particular persons in practical activities which permit these persons to live freely and rationally. The crucial points are two. First, though supra-individual, they are, like the moral law, structures within which the individual person actualizes his freedom. Unlike morality, they are objective institutions. They are nonetheless compatible with the free individual because the individual can understand each of them as actualizations of practical reason, hence as expressions of aims to which he himself can – and, unless he is an eremitic monk, in fact does – subscribe. The second point is that they are structures which are not only compatible with free individuals. They are, Hegel thinks, necessary to understand individual persons as really free, to exercise their will in property and contract (which, we recall, implicitly requires law, civil and criminal), and to live morally.

These supra-individual structures are the family, civil society, and the state. The family is that social structure in which freedom is substantial and actual but not reflectively rational. Family members, at least as much by instinct as by reflection, act for the good of the family, and children are reared to become independent members of society. Civil and economic society, on the other hand, is explicitly rational, but also explicitly self-interested, or subjective. It is a whole of self-interested parts whose interests overlap. The objective rationale rendering society a whole consists of the coordinating mechanisms of the market and of the legal system and regulatory and enforcement procedures to maintain the system. The constitutional state gives civil and economic society a framework that is more than accidental. In the state alone is freedom, the actualization of free will, understood as both universal, or rational, and objective. This understanding has two facets: on the one hand, the legislative, the executive, the judiciary, and the laws and regulations and bureaucracy are supposed to be occupied with a common good superior to that of each of the parts and of their coordination. (Doubtless, each of these branches of government – and especially the legislature – reflects and to some extent furthers, the interests of particular parts of society; but each is supposed to effect a common good which does not boil down to that of some part or of the mere coordination of the parts.) The second facet, not appreciated by Hegel,[13] is that the particular governmental structure is supposed to engage the citizens in self-reflection upon pursuits of their common good.

The importance of Hegel's *Philosophy of Right* for understanding Kant is that it gives the framework necessary to understand why the universal moral law, i.e., acting out of respect for the categorical imperative, is compatible with the individual's freedom. Kant says that morality must be compatible in order to understand how we can be both free and categorically obligated. But Hegel shows that morality must presuppose a stage at which the individual can realize his own will, or claims of right, understand clashes between his claims and others', and so understand the distinction between right in itself and particular claims of right. Morality is what is right in itself and rational. Hegel also explains in what sense what is right and rational is objectively actual, not at the personal but at the institutional level.

Our discussion of the categorical imperative as criterion – which contradicts Hegel's criticism of Kant – shows how the moral law is realized at the personal level. However, Kantian morality assumes the harmony of the kingdom of ends, i.e., the conformity of different moral subjects' universalization of their maxims. Such an assumption cannot account for actual social structures in which individuals have concrete rights and duties. Hegel gives this account, integrating Kantian morality into what is necessary to understand its relation to actual communities and social institutions.

B. *Engelhardt's Adaptation of Hegel*

Engelhardt makes two very important uses of Hegel. First, he uses Hegel's criticism of Kant to reinforce his own view that moral theory cannot ground content. Hegel argues that mere universalization, apart from extrinsic considerations, cannot specify moral content (##134–135). We agree; but both the notion of rationally willing and of the moral subject as an end in itself does allow some specification. Second, and more importantly, Engelhardt construes Hegel's concept of the state as encompassing a variety of content-full moral visions but, with respect to its moral authority, able to embrace and enforce none of them. In this sense, Hegel according to Engelhardt breaks with the notion of the state, which he identifies with Aristotle but thinks has dominated Western political thought at least since the Enlightenment, as authorized to structure society in terms of one or the other "single moral narrative" (Engelhardt, 1994, pp. 218 *et seq.*). "Hegel makes the watershed contribution of breaking with the Aristotelian ideal of polity. . . . The state as the self-aware political unity of diversity must be neutral regarding the content-full communities it compasses. It is ideally a limited democracy" (Engelhardt, 1994, p. 220). The emphasis is on "limited" rather than on "democracy," for if the representatives of a majority were to legislate a single-tier healthcare system, Engelhardt would say that this is the imposition of a content-filled moral view upon the unconsenting minority. (It may be presumed that the minority, even if it consists of one, has not given explicit, nor clearly implicit, consent to anything legislated by the majority. Hence, for Engelhardt, individuals within the state retain the right of the black market, to dispose of their resources with consenting others as they see fit, so long as unconsenting innocent others are not harmed thereby.)

> Since the state is not a larger-scale community, it cannot give the moral content and direction afforded by membership in a particular content-full moral community. Indeed, to seek it here would be a category mistake. On the other hand, it provides a categorial unity that is a *novum*. The state provides a disinterested administration of laws and procedures in terms of a new general identity as citizens (Engelhardt, 1994, p. 221).

> Contrary to Hegel's expectations, this social category is best understood as a neutral limited democracy. . . . A state that suppresses the content-full, peaceable diversity of its communities evacuates its specific content . . .

> (In a note he adds, "Such a state would not simply be a deficient realization of the state, it also would be tyrannical and repressive" [p. 223 n. 16].)

> "Even more central than the democratic character of the state, for which Hegel argued in categorial terms, is the state's character as a limited democracy" (Engelhardt, 1994, p. 222).

Engelhardt obviously thinks that the sole positive tasks of the state are those which are licensed by the governed, by each and all of them. Otherwise it unjustifiably uses its power and the resources it takes from the governed to impose one particular point of view on those who peaceably dissent. Engelhardt's understanding prevents the imposition of any political party's program, e.g., regarding health care, were the party to control the government in a representative democracy. For any such program would require taxes, and hence the redistribution of resources imposed by law and backed by force, with which it may be presumed some – e.g., libertarians – do not consent. Plausibly, such an understanding would prevent taxes for streets, sanitation, public health, or public education, and even for the common defense.

The problem with this view is that gives no content to the "new general identity [of those governed in a representative democracy] to be citizens." There is on its face no reason that anyone should consent to the use of his resources in a way that does not conduce to his, or his community's, long-term advantage. There is no common weal which marks Aristotle's notion of the polis as something to be distinguished from economic society of the village. The "disinterested administration of laws and procedures" could as well, and even better because more effectively, be undertaken by independent contractors.

Such a view renders all secular pluralist representative democracies illegitimate, for all depend upon a tax base that is not subject to veto by the self-interested desires of its constituent voters. (If, on the other hand, taxes were subject to veto by the voters *qua* citizens, the ground for veto would be an appeal to voters as political, an appeal that carries with it a *prima facie* obligation not to opt out of but to reform the body politic, and this introduces political stability.)

Why does he hold such a view? Because there is no justification of state authority save "the actual consent of the citizens to the actions of the government" (Engelhardt, 1996, p. 171; cf. pp. 167ff.) Appeal to a particular notion of God cannot ground a government of secular pluralist societies. A "morally canonical concrete understanding of the good life and moral obligations," he has already argued, cannot be justified by sound argument. Attempts to justify governmental authority by hypothetical contract are ineffective, because they "must impute to the hypothetical contractors a particular moral sense" in order to rank some social goods necessary to the contract (Engelhardt, 1996, p. 168). An actual past original agreement does not bind those who did not actually agree. More intriguingly, justifications based on prudence, that government authority prevents worse mayhem than the lack of such authority, are coercive, and thus not free agreements (Engelhardt, 1996, pp. 169–170). "[T]he final source for the moral authority of government [is] the actual consent of the citizens to the actions of the government. . . . To become a citizen is to have a share in a limited corporate undertaking [in which u]nconsented force can be forbidden, contracts enforced, and health care provided as a welfare right out of commonly held resources. It is simply that such secular moral authority will not extend to the control of the consensual action of free individuals, including their use of their private property" (Engelhardt, 1996, p. 171).

But the only reason individuals have to consent to the use of their resources to enforce contracts and forbid unconsented force is that it is in their long-term self-interest. And if they are to consider such from the standpoint of themselves as individuals, not as citizens committed *ab initio* to the state, then they will not freely give the government taxing powers via representative democracies.

The crux of Engelhardt's argument is the use of morality to justify social institutions covering pluralities. But morality is extra-institutional: it is addressed to the individual, *vis-à-vis* any other, not to particular groups of individuals. As Hegel has shown – and I hope I have indicated above – the viewpoint of asocial or extra-institutional morality construes the universal not as concrete, only as ideal. To assume that the moral viewpoint should be normative for the concept of the state, the political and legal structure which claims a monopoly on force and ensures a peaceable community among moral strangers, is a category mistake. The problems that result become clear on closer examination of the state.

C. *An Evaluation of Engelhardt's Concept of an Enforceable, Peaceable Community of Moral Strangers*[14]

Engelhardt would have the state legitimated by consent of the parties – individuals and communities of individuals, together with their resources – which constitute it, each acting out of self-interest. I want to argue, by contrast, that one must understand the state to depend upon the consent of the governed insofar as each recognizes that its enlightened self-interests requires a notion of a common good that cannot be understood in invisible-hand terms, as the uncoordinated result of the self-interested pursuits of the affected parties. Rather, the self-interest has to recognize the advantages of being a citizen, being political by nature.

The state for Engelhardt institutionalizes those functions of morality – we want to say justice – which are valid apart from political constitutions but need states for their realization. These are functions such as the curbing of unconsented force, the enforcement of contracts, provision of welfare from commonly held resources, but also necessary functions for which there is not sufficient commonality of understanding. In these cases, if market mechanisms, the result of numerous acts of consent by individuals without a common moral vision, do not work, then "robustly limited democratic mechanisms," limited only to clarifying points of controversies, are to be employed. In no case are "the consensual actions of free individuals" to be controlled, including their use of their private property (Engelhardt, 1996, pp. 171–172). The crucial point for Engelhardt is the moral one that "people own themselves as well as things privately" (Engelhardt, 1996, p. 396), apart from states. In positively delineating conflicting claims to property, or drawing common boundaries, etc., the state is simply an efficient reflection of this moral point.[15]

To the contrary, I want to say, this resolution of conflicting claims of right makes no sense apart from public legal authority (i.e., claims to represent the common good), enforce decisions and penalize cheaters. One cannot explain the mutual recognition of private property, or the adjudication of contracts and the elimination of cheaters – one cannot explain the peaceable society of moral strangers – unless one presupposes a public legal authority.

Consider that rational self-interest leads to the defense of one's rights by a protective association which is dominant within a given territory, that this association must prevent non-members from resorting to their own adjudication of conflicts with its members (e.g., oneself), and that it must morally compensate these non-members for preventing them from their own right of self-defense by giving them protection, even though they have not paid for it. This is the way Nozick (1974, pp. 305–308, 312–313) explains the development of the ultra-minimal state and the transition from it to the minimal, and alone morally justifiable, state. I ascribe this explanation to Engelhardt because he subscribes to Nozick's state, indeed, claims to give it the argumentative ground that it needs to constitute "the necessary conditions for the possibility of a moral structure" – i.e., the only morally authorized structure – "that members of morally diverse communities can, as moral strangers, recognize as mutually binding" (Engelhardt, 1996, p. 396). As Höffe points out (1995, p. 299; cf. p. 298), however, "it is rational to ignore such moral obligations so long as there is no public, i.e., non-partisan (in principle) coercive institution of enforcement." Unless the coercive authority is committed to the common good to begin with, there is no reason that it should not cheat. Institutions like the state are necessary because private, self-interested individuals cannot be expected not to cheat. The state must, then, be created to nurture a common good as a public interest. It is an interest that the various citizens have, because it is necessary to preserve their rights to life, limb, and property and to have just adjudication of disputes. That is to say, it is necessary as a public authority to settle disputes and draw boundaries, and it is necessary as a public legal authority capable of penalizing and coercing cheaters.[16]

Such as state can be construed as an ideal social contract (Höffe, 1995, p. 297) in which citizens both renounce freedom, e.g., to settle differences about property boundaries or contractual terms by force, and concede legitimate authority as necessary for the state to enforce settlement as well as to penalize cheaters. The citizens can be viewed as recognizing state authority as legitimate only if it does these things. Such a state would be democratic in order to reflect the consent of the governed. It would constitutionally preserve its rights. It would have an independent judiciary.

The crucial point of difference from Engelhardt's state is the state here sketched can only work if it can claim to represent a common good which is distinct from the uncoordinated result of self-interested pursuits. The sense in which the citizens have to consent to its function must presume that they are interested in judging the common good. One might still claim that such a state should be a minimal state, and should have no positive social functions not explicitly agreed upon by all citizens touched by those functions. But if it must serve the common good, it cannot do so if its citizens only have a sense of their own private good. It seems that the state would have to further what is necessary to realize senses of the common good, such as functioning so as to insure highways and efficient movement of traffic and communications, sanitation, public health, literacy, or promoting decentralization and privatization of public functions, and enforcing laws against economic cartels, or the more positive steps of encouraging public virtue, such as encouraging informed voter participation. These are the sorts of steps which, on the one hand, seem necessary to have a sense of the public good and, on the other, require more than is permitted Engelhardt's minimal libertarian state.

V. AFTERWORD

The body of this paper has been taken up in an analysis of Engelhardt's critique of moral theory and sketch of what is necessary to exercise authoritative actions which use moral strangers. I have argued that Kant's transcendental moral grounding is more fundamental than Engelhardt's, and that Hegel's analysis of it shows that we can understand why morality is a personal ideal only on the background of relations of property and contract which themselves presuppose the law of a more-than-libertarian state. My arguments have been large, and imprecise. But they do have the virtue of trying to engage Engelhardt seriously in his adaptations of Kant on morality and Hegel on society and the state. They also have this vice, among others, of not having the benefit of having been discussed with Stuart Spicker before being set to paper. I hope that they will entertain Stuart, and that their author will yet benefit from his comments.

Holy Archangels Greek Orthodox Monastery
Kendalia, Texas,
U.S.A.

NOTES

[1] Engelhardt shows how the major appeals in fact beg the question. First, he points out that one cannot rely upon basic intuitions, because these differ even about the content and the interpretation of foundational premises and principles. Moreover, they can differ widely enough so as to make the resolution of ethical controversies by casuistry impossible. The differences about moral standards can be basic (Engelhardt, 1996, pp. 42–45). Such differences are obvious in the case of morally judging nontherapeutic abortions; but they even bear on the analysis of such "taken-for-granted rules as 'It is immoral to torture or to kill the unconsenting innocent [i.e., those who have not used other moral agents without their permission] without their consent'" (cf. Engelhardt, 1996, p. 37). Second, Engelhardt points out that comparison of the consequences of different systems of moral choices cannot provide a defense of the foundation of one's moral theory (Engelhardt, 1996, pp. 45–48), because the comparison presupposes a standard by which different sets of consequences are to be rendered into a common denominator and judged. Such a standard will itself require defense. His criticism is sound. It is notable that he examines consequentialism on its own terms, disregarding entirely questions of the worth of each individual relevantly affected by the consequences, as well as the question of the intrinsic moral relevance of consequences. Third, Engelhardt turns to attempts to defend moral theory in terms of analyses of the structure of the moral reasoning or of rational choice (Engelhardt, 1996, pp. 49–53). These attempts are characterized by self-reflexivity and cover a variety of strategies: hypothetical-choice theory, whether the chooser be one or many, as with Rawls' hypothetical contractors, as well as theories of rationality, neutrality, or impartiality. The generic defect is that such attempts are never neutral; they have a particular sense of moral sense, or a particular "thin theory" of the good. They must incorporate some standard for moral content, otherwise they would not permit a defense of one's first premise(s). If, for example, a theory of rationality did not incorporate such a standard, it could not help resolve moral controversies or defend one's moral premises. The problem is that there can be basic differences about what constitutes such a standard, as in the contrasting views of Nozick and Rawls. (For the sake of completeness, one should note that he also considers Habermas's discourse theory [Engelhardt, 1996, pp. 53–55] and game theory [Engelhardt, 1996, pp. 55–56].)

[2] Because the goal of moral action is achieving goods and avoiding harms, morality for Engelhardt requires, besides the principle of permission, also the principle of beneficence (cf. Engelhardt, 1996, pp. 123–124). Given the failure of moral theory to ground a canonical account of goods and harms, he construes a content-filled principle of beneficence as that to which a community assents.

[3] For a brief but nuanced discussion of the difference between ethics addressed to the individual moral agent and that addressed to him in his social roles, see Hartmann, 1981, ch. 1, to whom my discussion in this paragraph is indebted.

[4] There is general property, in the sense that all have a right to the world's land and material which is the raw stuff taken possession of and reworked into what particular persons' own, their property; so all must be equally compensated by owners who, in taking portions of the common land and material, restrict others' opportunity to transform it into their own property (Engelhardt, 1996, p. 161; cf. pp. 158–159). But the concept of general property may be disregarded within the scope of this discussion.

[5] This argument is defeated, if personal property (and its contractual exchange) is a right which is not practically realizable apart from a rule of law. For then one could talk about an ideal contract between state and individual, whereby each individual gets the right to possess and use and exchange personal property by relinquishing the claims to others' life, limb, and property and to the state the authority to enforce what is necessary for protection of life, limb, property, and contract. This is the tack I shall take below.

[6] Though I have used "ethics" and "morality," and "ethical" and "moral" interchangeably, ethics is, properly speaking, theoretically based morality.

[7] My presentation of Kant's ethics is much indebted to (what I understand of) Aschenberg, 1996.

[8] Development of one's own talents or help of the needy is a meritorious rather than a strict duty, because it can only be defended at times, not at all times.

[9] The addition of the adverb is meant to call attention to the distinction between wishing and willing, as well as that between what moral subjects would will and what one would be willing to undergo if the situations were reversed, e.g., and I were the crippled beggar instead of the possible donor.

[10] One can justify Kant's defense of the meritorious duties not to neglect one's talents or to disregard contributing to the needy. One can conceive of oneself not unfailingly striving to develop one's talents, or of a society in which persons often used surplus funds to enjoy themselves rather than to help the needy. However, to neglect the needy altogether is (*in. al.*) to neglect the real conditions that encourage them to act morally: it is (*in al.*) to deny that we have a responsibility to encourage (positively) moral conduct in others. Analogously, to neglect my talents in general is to neglect conditions that encourage (positively) moral conduct in myself.

I must, therefore, disagree with Engelhardt's distinction of a conceptual contradiction from a contradiction in will. In fact one cannot conceive – or will – a "notion of the peaceable community, in the sense of a community whose authority is not based on force" unless one allows that some conditions for it encourage the responsible use of persons' freedom and others do not. This, however, is to take us to the second question with regard to Engelhardt, the real conditions of a morally compatible – "peaceable" – community.

[11] An interesting question, which the constraints of this paper preclude pursuing, is: Who are the relevant individuals whose consent is required, all voting citizens? all citizens who are moral agents? all individuals whose taxes support government? There is also the correlative question, whether for Engelhardt those who pay more taxes should have more say in government.

[12] Ethical life for Hegel is the synthesis of self-conscious subjectivity [which, apart from the objective moral good is either one's private conscience or moral evil], and objectively realized right [which, apart from moral subjectivity, is manifested as wrong]. Ethical life is "the concept of freedom developed into the existing world and the nature of self-consciousness" (#142).

[13] The state is distinguished from civil society as being more than an optional association formed to protect individual interests (#258). The crucial question with regards to Hegel's articulation of the state in its internal constitution (#261ff.) is to understand in what respect the "state [as Hegel has it] is the actuality of concrete freedom," and in

what respect Hegel's articulation can accommodate what we want of a state, that it is the real basis of the freedom of its citizens. The problem is that he gives conflicting articulations of the internal constitution of the state: at #273 into legislature (which would be the stage at which the classes of civil society mediated their class interest with their universal interests as political, i.e., as citizens), executive, and the crown (or embodiment of the state *vis-à-vis* other states), and the execution, into crown first (#275), then the executive (populated by the civil service, itself a social class (cf. #297, a factual truth)), and the legislative, simply mediating the commands from the top. Hegel is very concerned about social atomism (cf. Hegel, 1967, p. 175 (against the division of powers), 178, 182–183, 195f.); his problem, however, is that he views the individuals governed as primarily social, not (also) political, or citizens, concerned with a common supra-social good.

[14] The following arguments are greatly indebted to Höffe.

[15] "The state is precisely that social structure in which one can understand a moral authority binding persons and communities with different moral visions and through which general rights to forbearance can be secured and protected. The state is a way of achieving social organization that is not simply another community". The Enlightenment hope "to discover in reason a canonical basis for authoritative, large-scale, secular communities . . . has been shown to be vain. The state is rather a social structure that can compass a diversity of communities protecting both the rights of those communities and the individuals who are its members" (Engelhardt, 1996, p. 177).

[16] I pass over the necessity of institutions of the state that are necessary to guarantee justice that is commutative but diachronic, most notably, of parents to their children and of grown children to their aged parents (cf. Höffe, 1995, pp. 279–281). Engelhardt would perhaps dispute this point.

BIBLIOGRAPHY

Aschenberg, Reinhold: 1996, 'Warum sollen wir moralisch gut sein? Bemerkungen zu einigen Fragen der Ethik' (manuscript of lecture given in February 1996 to religion teachers at gymnasia), Tübingen.

Engelhardt, H. T., Jr.: 1996, *Foundations of Bioethics* (2nd edition), Oxford University Press, New York.

Engelhardt, H. T., Jr.: 1994, '*Sittlichkeit* and Post-Modernity: A Hegelian Reconsideration of the State', in H. T. Engelhardt Jr. and T. Pinkard (eds.), *Hegel Reconsidered. Beyond Metaphysics and the Authoritarian State*, Kluwer Academic Publishers, Dordrecht.

Hartmann, K.: 1981, *Politische Philosophie*, Verlag Karl Alber, Munich.

ten Have, H. A. M. J.: 1997, 'Bodies of Knowledge, Philosophical Anthropology, and Philosophy of Medicine', in this volume.

Hegel, G. W. F.: 1967, *Hegel's Philosophy of Right*, T. M. Knox (trans. and notes), Oxford University Press, New York.

Höffe, O.: 1995, *Political Justice*, Polity Press, Cambridge.

Kant, I.: 1956, *Critique of Practical Reason*, H. J. Paton (trans.), Bobbs-Merrill Company, Indianapolis; also cited according to *Kants Werke*, Akademie-Textausgabe , Band V (AK V), Walter de Gruyter & Co., Berlin, 1968.

Kant, I.: 1964, *Groundwork of the Metaphysic of Morals*, H. J. Paton (trans. and notes), Harper & Row, New York; also cited according to *Kants Werke*, Akademie-Textausgabe, Band IV (AK IV), Walter de Gruyter & Co., Berlin, 1968.

Kant, I.: 1991, *The Metaphysics of Morals*, M. Gregor (trans.), Cambridge University Press, Cambridge; also cited according to *Kants Werke*, Akademie-Textausgabe, Band VI (AK VI), Walter de Gruyter & Co., Berlin, 1968.

Nozick, R.: 1974, *Anarchy, State and Utopia*, Basic Books, New York.

Spicker, S. F.: 1975, 'The Lived-Body as Catalytic Agent: Reaction at the Interface of Medicine and Philosophy', in H. T. Engelhardt Jr. and S. F. Spicker (eds.), *Evaluation and Explanation in the Biomedical Sciences*, D. Reidel Publishing Company, Dordrecht, pp. 181–204.

Spicker, S. F.: 1990, 'Invulnerability and Medicine's "Promise" of Immortality: Changing Images of the Human Body During the Growth of Medical Knowledge', in H. A. M. J. ten Have, G. K. Kimsma, and S. F. Spicker (eds.), *The Growth of Medical Knowledge*, Kluwer Academic Publishers, Dordrecht, the Netherlands, pp. 163–175.

Wagner, H.: 1992, *Die Würde des Menschen*, Königshausen & Neumann, Würzburg.

H. TRISTRAM ENGELHARDT, JR.

BOLE ON KANT, HEGEL, AND ENGELHARDT: A BRIEF REPLY

Thomas J. Bole provides an ingenious set of fundamental issues that bear on the philosophy of medicine. In so doing, Bole further vindicates Stuart Spicker's conviction that bioethics involves substantial philosophical issues that require sustained analysis. Since Bole in his essay gives such attention to my work, he deserves at least a brief note in response. Though I find his arguments to be engaging, I also judge them to be misguided. They fly wide of the target of my arguments because of a cluster of critical misconceptions with which he begins. Most significantly, in his essay he confuses (1) the role to be played by a common framework of right-making authorizations and (2) the role that might be played by a common understanding of the good. Indeed, he does not appreciate the challenge confronting secular moral philosophy in the de facto and de jure moral diversity of our post-modern, post-Christian, post-traditional age. Secular morality must span real moral diversity through a framework of common authority. Had Bole attended to the role that could be played by a common right-making practice, a transcendental possibility for a common procedural framework of moral authority, he would have had at his disposal a quite different understanding of the "common will". He would have seen this approach as not only allowing a recasting and grounding of Nozick's *Anarchy, State, and Utopia* (Nozick, 1974), but also of the reflections provided by others such as the contemporary Italian liberal thinker, Bruno Leoni (1913–1967) (Leoni, 1991). Bole would then also have been able to appreciate why neither I nor Hegel would regard the state as simply a unity of individuals. The state is most properly a unity of individuals already in moral communities (as well as of individuals who fail to achieve a communal life and therefore have no unambiguous locus in a moral community). Consider in particular Bole's criticism of both Hegel and myself in footnote 13 of his essay. Here Bole commits a category mistake. He does not appreciate that the state is precisely that unity that spans a diversity of social structures. In footnote 10, Bole also fails to understand the force of Hegel's criticisms against Kant and therefore against Bole's attempt to secure Kant against my arguments. Bole

H. T. Engelhardt, Jr. (ed.), The Philosophy of Medicine, 223–225.

fails to recognize that there is not one secular moral view of talents or of the proper approach to human needs. The talents and virtues of Gauguin in Polynesia are not those of Kant in Prussia, nor those of a Samurai warrior. Nor will secular rationality by itself allow one to know whether autonomy is best schooled by beneficence or hardship.

Bole's Aristotelianism leads him to fail to appreciate that the cunning of reason allows persons, individuals, and communities to pursue diverse goals and still achieve coherent collaboration (Hegel, *The Lesser Logic*, § 209). The limited state, as that moral framework within which individuals can collaborate under conditions of drawing authority for the use of persons only from their permission, allows much to be done at a political level. This is especially the case when individuals are already members of robust communities. Such communities can out of diverse motivations, understandings of the good, and senses of prudence, collaborate in and through the state. For this to function fully, it is true that one must abandon the view of geographically located communities and revise the understanding of the nation-state as a society that is equivalent to a national community. This is far from justifying the state as merely an efficient mechanism, a point that Bole unwittingly concedes in footnote 10. The state provides a political unity spanning diverse communities and persons motivated by varying views of the good. As such, the state is a categorial novum.

Here is not the place to rehearse my position in detail, but only to remind Bole that, if he had consulted Kant in the *Metaphysik der Sitten* (AK VI 305–308, 312–313), he would have discovered an argument for why one must leave the state of nature for a political order, which bears many similarities with that offered by Nozick, though Kant's account possesses greater categorial depth. When Kant's argument is freed of the presumption that one can in general secular terms discover a canonical, content-full understanding of the good, this approach provides a basis for justifying an account of the state not unlike that which I have defended (Engelhardt, 1996). Though persons cannot be presumed to have consented to the range of activities undertaken by current large-scale states, they can as members of voluntary, non-geographically based communities contribute robustly to the common goods of a state that can span numerous such communities. Such states can provide the equivalent of welfare rights without sharing a common understanding of the good or of human flourishing. All this can occur without ascribing a special value to peace. Indeed, the secular "kingdom of ends" is not most appropriately described as the peaceable kingdom. Rather, it is that framework within which moral strangers can participate in a common moral world, though they do not share a common view of the good. What is offered by my account of the state is a social category within which there can be common secular, morally authoritative action in the face of real moral diversity.

BIBLIOGRAPHY

Engelhardt, H. T. Jr.: 1996, *The Foundations of Bioethics*, 2nd ed., Oxford University Press, New York.
Leoni, B.: 1991, *Freedom and the Law*, 3rd ed., Liberty Press, Indianapolis, Indiana.
Nozick, R.: 1974, *Anarchy, State and Utopia*, Basic Books, New York.

AKIO SAKAI

EPILOGUE: THE USE OF THE PAST[1]

It is easy, in a sense, to explain the characteristics of Spicker's thought, but at the same time it is extremely difficult. It is easy, because we can say that his philosophy consists of both a diachronic stream of ideas tracing back to ancient times, and a synchronic theoretical framework including analytical philosophy and Continental phenomenology. However, it is difficult to recapitulate the method and content of his thought from a single perspective, for his thought is unique, though he freely adopts a variety of notions from other philosophers.

A. MacIntyre pointed out the confusion that occurs between past philosophies and us, the contemporary readers.

> Either we read the philosophies of the past so as to make them relevant to our contemporary problems and enterprises, transmuting them as far as possible into what they would have been if they were part of present-day philosophy, and minimizing or ignoring or even on occasion misrepresenting that which refuses such transmutation because it is inextricably bound up with that in the past which makes it radically different from present-day philosophy; or instead we take great care to read them in their own terms, carefully preserving their idiosyncratic and specific character, so that they cannot emerge into the present except as a set of museum pieces (MacIntyre, 1986, p. 31).

MacIntyre asserts here that actually there are two ways to employ the past. I want to focus on the expression "a set of museum pieces" in a way that departs from MacIntyre's intention.

It is not unusual for a reference to ancient philosophy to appear in Spicker's essays. Through his quotations various types of episodes, as well as past philosophies, are included. They all share the characteristics of "museum piece." Importantly, the museum piece has a very unique and positive function for it is not a silent relic of the past playing no active role. In the following, I analyze his style as reflected in his essays (a number of them have been translated into Japanese).

For example, in 'Chronicity' there is a quotation from Plato that depicts, through a worker's way of living in ancient Greece, the resistance to paternalistic medical therapy, and it mocks a culture's preoccupation with life-prolongation (Spicker, 1982).

H. T. Engelhardt, Jr. (ed.), The Philosophy of Medicine, 227–234.

> If anyone prescribes for him a long course of treatment with swathings about the head and their accompaniments, he hastily says that he has no leisure to be sick, and that such a life of pre-occupation with his illness and neglect of the work that lies before him isn't worth living (Plato, *Republic* III, 406d–e).

This passage makes us realize that some important issues, like resistance to the "medicalism" or respect for a patient's autonomy were already described explicitly 2500 years ago; at the same time it makes us wonder what changes occurred in clinical medicine during this long period. Especially the latter provides us with the motive critically to consider the factors that served today's controversial clinical setting.

The term 'xenodochium' is used in the title of a brief essay that addresses American medical education (Spicker, 1975a). Consulting a Latin-Japanese dictionary, one finds that this term denotes (1) an inn for foreigners, and (2) a hospital. Spicker's point is clear: the temple of Asclepius as the site of medical education should serve also as the inn for Apollo, the symbol of philosophy and the other "medical humanities." However, this reference is subtle. We should recall that Asclepius is the son of Apollo. They became estranged from each other and therefore they could no longer live together. The Latin word 'xenodochium' casts light not only on the contemporary problem of the dissociation between medicine and the humanities, but also on the important task that still remains for medical schools; to provide a "guest suite" for the humanities.

The epigraph at the very beginning of Spicker's essay on surrogate motherhood criticizes this very concept by means of a quotation from Genesis (Spicker, 1990).

> When Rachel saw that she was not bearing children to Jacob, she became jealous of her sister [Lia] and said to Jacob, "Give me children or I shall die." Jacob became impatient with Rachel and answered her, "Can I take the place of God, who has made you barren?" She replied, "Here is my slave-girl, Bala; go in to her that she may bear on my knees [participate in the ceremony of adoption], and I too may have children by her." She gave him Bala, her maid, in marriage, and Jacob had relations with her. Bala conceived and bore Jacob a son. Then Rachel said, "God has pronounced judgment in my favor, for he has heard my prayer and given me a son" (Genesis 30, 1–6).

The content of the epigraph not only underscores an early account of collaborative reproduction, but also signals its essence, or at least the motive for this form of human reproduction. The basic problem is one of desire. In the epigraph, the extraordinary desire of Rachel is described;

she is desperate to have a child, even coopting her slave-girl. The preplanned "give and take" of an infant that involves a third party is successfully accomplished. They bless the baby as God's gift. Such joy and the desire motivating it signal that "agent-centered moral theory" is virtually neglected. The essay clarifies the fact that the true hero or heroine of "collaborative reproduction" is not the adoptive mother Rachel, nor the parents Jacob and Bala, but the child, who appears to play a non-speaking part in this human drama.

In discussing the notion of the lived-body, Spicker considers Plotinus (Spicker, 1975b). For Plotinus' "spiritualistic metaphysics" disregards (or, better, despises) the human body. However, the person who poses the more serious problem here is not Plotinus, but his intimate friend Eustochius of Alexandria. A dramatic end awaits us in this essay: though Eustochius "consecrated himself exclusively to Plotinus' system and became a veritable philosopher," we are informed in the closing sentence that Eustochius was Protinus' physician. Owing to Spicker's skillful construction, we become vividly aware of the true danger, i.e., the irony that even the specialist who examines the human body (e.g., the physician) is apt to tacitly adopt metaphysical assumptions regarding the human body. A spiritualistic metaphysics may have the power to fascinate; however, once we forget the significance of the lived body, we become too far removed from human reality. Spicker calls the adherents of spiritualistic metaphysics, including Eustochius, "veritable philosophers," but of course, his actual view is expressed by the use of quotation marks.

A medieval portrait painting plays a powerful role in an essay on the "biology of aging" (Spicker, 1984). Here Spicker reinstates the positive value of old age. The Canon George van der Peale (painted by Jan van Eyck in 1436) is highlighted. The reader is asked to pay attention to the Canon's "richly lined face," for "the lines of the face, that is, the full physiognomy of the face of elderly is not the sum of all facial wrinkles." Here, too, Samuel Butler and the Darwinians are criticized by Spicker, for they introduced an arbitrary and mistaken view of the person who exists here and now, that eventually leads to our devaluation of aging and the aged. For these thinkers, it is imperative simply to preserve the human species. Accordingly, for them, individuals who lose the ability to reproduce lose virtually everything. Despite the fact that some biologists are eager to erase the lines in our faces, the ruggedness of the face of van der Peale is intrinsic to the human condition; hence they signify the unique and precious feature that can never be reduced to a mere set of cells.

That Galileo gave lectures and wrote in Italian, not in the language of the medieval intellectuals (Latin) is introduced in order to consider the philosophical and ethical problems raised by genetic engineering and therapy (Spicker, 1996). Latin is analogous to the technical terms of modern genetics and the complicated notions they represent. Aside from genetics, every articulated academic specialty has its own technical language and hence becomes evermore unintelligible to lay people. Such a phenomenon – the Latinization of academic specialties – must be translated to be comprehensible, especially when it represents the knowledge on which the future of human beings depends. Galileo employed the everyday language of the artisans in order for them to be able to make the tools he wanted. In other words, he spoke in the vernacular – Italian. In short, geneticists must "translate" their technical terms for lay people. The aim is to prevent useful knowledge from exclusive possession and to utilize the advances of science for all mankind. The reason Galileo is mentioned, is to stress the importance of transmitting scientific knowledge to the public.

The late Gilles Deleuze commented, in 1965, on the thought of Nietzsche; "Nietzsche integrates two ways of expression, the aphorism and the poem to philosophy. These forms themselves imply a new image of the thinker and the thought" (Deleuze, 1965, p. 17).[2] We only need replace "the aphorism and poem" with "the citation," in order best to express an aspect of Spicker's thought.

He leads us not to the historical episode itself, but to the contemporary distortion as clearly revealed in contrast to such a moment in history, the museum piece. A museum piece expresses two values. The first is derived from differences of time and space, i.e., from the piece's oldness and exoticness. Owing to piece's entirely "different" origin from the origin of our contemporaries, it is prized enough to motivate us to go to the museum to enjoy them. Because the museum piece belongs to a totally different context from our own, it provokes a contrast with the everyday that is part of its value.

Its second value is more individual. Our positive values with respect to certain things in the past , as well as their cultural differences they represent, prompt us to preserve those things in museums. The artistic power incorporated in "the picture," the painter's values, our feeling of awe toward the most eminent philosophers, the realization of the great influence of Greek myths, the attention to the Bible as the foundation of the Judeo-Christian point of view, and the importance of social systems as products of civilization – all these things compel us to reflect. The kinds of value they embody have their bases in the unique quality of any great expression of the human spirit, whether an historical even, a great work of art, or an abstract idea. It is a quality which ultimately separates itself from us on the basis of inexpressible difference.

What is indicated here is "difference" in the post-modern sense, a sense which transverses the time, space, system, genre, and the individual quality of the product or event. It is this sense that Stuart Spicker's writings seek to highlight. The method he adopts is thoroughly to reconstruct the past as the whole of "difference," and then to focus on selected episodes or things (fragments of "difference"), following a clue of conceptual similarity. His focus upon this "difference" has an extraordinary power that suddenly creates a heterogeneous field of force in the context of discussion, and eventually provides a point of action that elevates the intensity of the writing. What motivates such intensity is the contrast which is implicit in the analogy.

By virtue of Spicker's style, we are called to attend, as the audience, to the various marvelous scenes that ancient philosopher indicated concern in the limitation of modern medicine; the past accommodation signifies the existence of a still unresolved problem; the master and disciple, philosopher and physician, present themselves as persons from whose poor example one can learn a great deal. We are warned of the dangerous desire that clouds one's reason; the medieval painting teaches us what we really are; the Italian mathematician-astronomer prefers not solely to possess knowledge. Spicker fixes his eyes on the reality, assigns the parts of the play to the actors (characters living in the past), and directs them. Spicker, like Nietzsche, is "a thinker who 'dramatizes' ideas, in other words, presents them as successive events at various levels of tension"[3] (Deleuze, 1965, p. 38).

As a playwright-director, Spicker makes historical persons and systems speak "his own" ideas, and enhances their character in terms of the *Zeitgeist* of their days. The contrast of past and present produces a unique irony that transcends time and space. That is, the peculiar "reversed composition," in which the past description and system can be made to critique the present situation, and historical persons state "Spicker's own" ideas. This style leads us to a different time and space and forces us to experience going to and returning from their world. However, his style has the surprising effect of fixing our eyes on reality and makes us see things more clearly. This can only be possible when comparison and contrast come into full play.

The philosophy of medicine owes an enormous debt to many disciplines. It must incessantly commit to the things and materials in different dimensions, in order critically to re-examine the present status of medical practice and medical theory. Spicker's aim is, of course, the realization of improved clinical practice. He directs the actors from the past and brings the present closer to what he dramatizes. As his methodology is based on universal concepts, i.e., similarity and difference, its strength and power is not culture-bound. It seems reasonable to suppose that this is one of the reason why his essays have had great influence on Japanese bioethics and philosophy of medicine.

A CLOSING WORD

Spicker has retained a close relationship with Japan. He has been invited to Japan five times by various Japanese institutions: the Ministry of Education, the Japanese Association for Philosophical and Ethical Research in Medicine, and other agencies. The themes of his lectures reflect his broad interests: for example, 'Western Visions of Man, Medicine, and Technology,' 'The Nurse's Commitment to the Patient's Autonomy,' 'Marketing Health Care,' 'Humanistic Studies and Biomedical Ethics in U.S. Medical Schools,' 'Natural Death and High-Tech Medicine,' 'Invulnerability and Medicine's "Promise" of Immortality,' 'Collaborative Reproduction,' 'Chronicity and the Clinic,' and many others. These lectures cover most of the important problems and issues that today's physicians are forced to confront. The contents of these lectures have appeared in various academic publications and they continue to exert an important influence on Japanese scholarship.

For example, Spicker's lectures on the case of Baby M in the U.S. in 1989, which was later published in the *Annals of the Japanese Association for Philosophical and Ethical Research in Medicine* (Spicker, 1990), gave direction to Japanese scholars fully to understand the bioethical implications of this subject. Moreover, his lecture to a group of Japanese psychiatrists in 1991, concerning the meaning of time in chronic diseases from the phenomenological perspective, served to initiate the Japanese interest in indispensable psychological care for chronic patients. I believe that Spicker's activities in Japan will continue to encourage the study of bioethics and the philosophy of medicine.

In 1995, *Igakutetsugaku eno Shotai [An Invitation to the Philosophy of Medicine: The Collected Papers of Stuart Spicker]*, translated by R. Ishiwata, A. Sakai, and H. Fujiwara, was published in Japan. This book surely provided the Japanese with a more precise understanding of the philosophy of medicine and bioethics. Our hope is that he will continue to provide us with his unique perspective and stimulating philosophical analyses. Of course, I do not personally worry about about whether he will do so, because I am presented with new intellectual challenges (not to mention the fruits of his humor) each time I receive one of his letters and publications.

School of Medicine
Iwate Medical University
Morioka, Japan

NOTES

[1] Between 1988 and 1990, I was engaged in research covering philosophical, medical and historical perspectives under Professor Spicker's supervision, and at the same time in translation of his works, while visiting the Department of Community Medicine and Health Care, University of Connecticut, School of Medicine in Farmington. The most impressive thing for me, during my two-year stay, was the way his ideas and writings transversed academic disciplines and the standard bounds of bioethics taken in a narrow sense, though they were clearly grounded in philosophy, bioethics, and medicine. I concluded that his works should be regarded as a contribution to "philosophy of medicine" in its most comprehensive sense, for they were rooted in the two traditional disciplines, i.e., philosophy and medicine, but they also reflected other disciplines. I owe much to him, not only my abiding commitment to research in philosophy of medicine itself, but also in the basic methodology required to pursue academic research in general.

[2] Original text reads: "Nietzsche intègre à la philosophie deux moyens d'expression, l'aphorisme et le poème. Ces formes mêmes impliquent une nouvelle conception de la philosophie, une nouvelle image du penseur et de la pensée."

[3] Original text reads: "Nietzsche est un penseur qui <dramatise> les Idées, c'est-à-dire qui les présente comme des événements successifs, à des niveaux divers de tension."

BIBLIOGRAPHY

Deleuze, G.: 1965, *Nietzsche*, Presses Universitaires de France, Paris.
MacIntyre, A.: 1986, 'The Relationship of Philosophy to Its Past', in R. Rorty, J. B.

Schneewind, and Q. Skinner (eds.), *Philosophy in History*, Cambridge University Press, Cambridge, pp. 31–48.

Spicker, S. F.: 1975a, 'A Xenodochium for Apollo in the Temple of Asclepius: Medical Ethics Education in the U.S.A. – Status Quo and Future Task', *Connecticut Medicine* 39:1, 41–44. An expanded version was translated into Japanese by Akio Sakai, 'Asclepius shinden ni okeru Apollo no yado – America no igakurinri kyoiku: sono genjo to shorai no kadai', in [*Igakutetsugaku Igakurinri*] *Annals of the Japanese Association for Philosophical and Ethical Research in Medicine* No. 7 (1989), 113–122.

Spicker, S. F.: 1975b, 'The Lived-Body as Catalytic Agent: Reaction at the Interface of Medicine and Philosophy', in H. T. Engelhardt, Jr., and S. F. Spicker (eds.), *Evaluation and Explanation in the Biomedical Sciences*, D. Reidel Publishing, Dordrecht, pp. 181–204. Translated into Japanese (see Spicker, 1991) by R. Ishiwata, A. Sakai and H. Fujiwara, 'Baitai toshiteno ikirareru shintai: igaku to tetsugaku no setten niokeru hanno', in *Igakutetsugaku eno Shotai*, Jiku shuppan, Tokyo, 1995.

Spicker, S. F.: 1982, 'Chronicity: Some Definitional Issue', in *Newsletter: Committee on Philosophy and Medicine* (American Philosophical Association) 13, 2-32-3. An expanded version was translated into Japanese by Akio Sakai (see Spicker, 1991, 'Mansei no gainen to sono rinshoteki imi') and published in [*Ijigaku Kenkyu*] *Journal of Interdisciplinary Humanistic Medical Research* 6, 153–167, Iwate Medical University, Morioka, Japan, 1991.

Spicker, S. F.: 1984, 'Philosophical Reflections on the "Biology of Aging"', in S. F. Spicker and S. Ingman (eds.), *Vitalizing Long-Term Care*, Springer Publishing Company, New York, pp. 29–45. Translated into Japanese by R. Ishiwata, A. Sakai and H. Fujiwara, '"Roka no seibutsugaku" ni tsuiteno tetsugakuteki kosatsu', in *Igakutetsugaku eno Shotai*, Jiku shuppan, Tokyo, 1995.

Spicker, S. F.: 1990, 'Collaborative Reproduction: Legal and Ethical Aspects [The Case of "Baby M" in New Jersey Courts]' 'Tasha no kyoryoku niyoru seishoku-keikaku: hoteki rinriteki kanten karanokosatsu', translated by Akio Sakai [*Igakutetsugaku Igakurinri*] *Annals of the Japanese Association for Philosophical and Ethical Research in Medicine* No. 8, 90–101.

Spicker, S. F.: 1996, 'The Unknowable Effects of Genetic Interventions of Future Generations (Or, Who Guards the Genetic Engineers in Democratic Republics?)', in E. Agius and S. Busuttil (eds.), *Genetic Engineering and Obligations to Future Generations*, Kluwer Academic Publishers, Dordrecht. An edited, shorter version is translated into Japanese by Akio Sakai, 'Idenshi sosa no shirarezaru eikyo to miraisedai,' [*Seminar Iryo to Shakai*], *Journal of Health Care, Medicine and Community* No. 9, 28–36.

BIBLIOGRAPHY OF THE WORKS OF STUART FRANCIS SPICKER

BOOK

(1995) *An Invitation to the Philosophy of Medicine* (*Igaku-tetsugaku e no jyosho*), Ryuji Ishiwata, Akio Sakai and Hiroshi Fujiwara, translators, Jiku Publishers, Tokyo, Japan.

EDITED VOLUMES

(1970) *The Philosophy of the Body: Reflections of Cartesian Dualism*, Quadrangle Books, Inc., Chicago, reprinted, New York Times Book Company, New York.

(1975) *Evaluation and Explanation in the Biomedical Sciences*, with H. T. Engelhardt, Jr. (co-editor), Kluwer Academic Publisher, Dordrecht.

(1976) *Philosophical Dimensions of the Neuro-Medical Sciences*, with H. T. Engelhardt, Jr. (co-editor), Kluwer Academic Publishers, Dordrecht.

(1977) *Philosophical Medical Ethics: Its Nature and Significance*, with H. T. Engelhardt, Jr. (co-editor), Kluwer Academic Publishers, Dordrecht.

(1978) *Aging and the Elderly: Humanistic Perspectives in Gerontology*, with K. M. Woodward and D. D. van Tassel (co-editors), Humanities Press, Atlantic Highlands, New Jersey.

(1978) *Mental Health: Philosophical Perspectives*, with H. T. Engelhardt, Jr. (co-editor), Kluwer Academic Publishers, Dordrecht.

(1978) *Organism, Medicine, and Metaphysics: Essays in Honor of Hans Jonas on His 75th Birthday*, Kluwer Academic Publishers, Dordrecht.

(1979) *Clinical Judgment: A Critical Appraisal*, with H. T. Engelhardt, Jr. (co-editor), Kluwer Academic Publishers, Dordrecht.

(1980) *Nursing: Images and Ideals (Opening Dialogue with the Humanities)*, with S. Gadow (co-editor), Springer Publishing Company, New York.

(1981) *The Law-Medicine Relation: A Philosophical Exploration*, with J. M. Healey and H. T. Engelhardt, Jr. (co-editors), Kluwer Academic Publishers, Dordrecht.

(1983) *New Knowledge in the Biomedical Sciences: Some Moral Implications of Its Acquisition, Possession and Use*, with W. B. Bondeson, H. T. Engelhardt, Jr. and J. White (co-editors), Kluwer Academic Publishers, Dordrecht.

(1983) *Abortion and the Status of the Fetus*, with W. B. Bondeson and H. T. Engelhardt, Jr. (co-editors), Kluwer Academic Publishers, Dordrecht.

(1984) *Vitalizing Long-Term Care: The Teaching Nursing Home*, with S. R. Ingman (co-editor), Springer Publishing Company, New York.

(1987) *Euthanasia and the Newborn: Conflicts Regarding Saving Lives*, with R. C. McMillan and H. T. Engelhardt, Jr. (co-editors), Kluwer Academic Publishers, Dordrecht.

H. T. Engelhardt, Jr. (ed.), The Philosophy of Medicine, 235–240.

(1987) *Ethical Dimensions of Geriatric Care: Value Conflicts for the 21st Century*, with S. R. Ingman and I. R. Lawson (co-editors), Kluwer Academic Publishers, Dordrecht.
(1987) *The Contraceptive Ethoss: Reproductive Rights and Responsibilities*, with W. B. Bondeson and H. T. Engelhardt, Jr. (co-editors), Kluwer Academic Publishers, Dordrecht.
(1988) *The Use of Human Beings in Research: With Special Reference to Clinical Trials*, with I. Alon, A. de Vries, and H. T. Engelhardt, Jr. (co-editors), Kluwer Academic Publishers, Dordrecht.
(1990) *The Growth of Medical Knowledge*, with H. ten Have and G. Kimsma (co-editors), Kluwer Academic Publishers, Dordrecht.

ARTICLES AND BOOK CHAPTERS

(1970) 'Report on the '"Philosophers Into Europe" Conference', *Journal of the British Society for Phenomenology* 1(3), 67–77.
(1970) 'Shadworth Hodgson's Reduction as the Anticipation of Husserl's Phenomenological Psychology', *Journal of the British Society of Phenomenology* 1(2), 57–73.
(1970) 'Comments', in E. Straus and R. Griffith (eds.), *Phenomenology of Memory: The Third Lexington Conference on Pure and Applied Phenomenology*, Duquesne University Press, Pittsburgh.
(1971) 'Rejoinder to Bruce Wilshire', *Journal of the British Society for Phenomenology* 2(3), 80.
(1973) 'Inner Time and Lived Time: Husserl and Merleau-Ponty', *Journal of the British Society for Phenomenology* 4(3), 325–347.
(1973) 'The Fundamental Constituents of Consciousness: Process-Contents and the *Erlebmisstrom*', *Man and World* 6(1), 26–43.
(1975) 'The Lived-Body as Catalytic Agent: Reaction at the Interface of Medicine and Philosophy', in H. T. Engelhardt, Jr., and S. F. Spicker (eds.), *Evaluation and Explanation in the Biomedical Sciences*, Kluwer Academic Publishers, Dordrecht, pp. 181–204.
(1975) 'A *Xenodochium* for Apollo in the Temple of Asclepias', *Connecticut Medicine* 39(1), 41–44.
(1975) 'The Language of PSRO: Bewitchment, Bureaucracy, and Biology', in I. A. Lawson and S. R. Ingman (eds.), *The Language of Geriatric Care: Implications for Professional Review*, Connecticut State Research Series, North Haven, Connecticut, pp. 58–67.
(1975) 'Inquiry and Commentary', in *Experiments and Research with Humans: Values in Conflict*, National Academy of Sciences, Washington, D.C., pp. 110–111, 145.
(1976) 'The Spurious Psyche: Some Distinctions', in S. F. Spicker and H. T. Engelhardt, Jr., (eds.), *Philosophical Dimensions of the Neuro-Medical Sciences*, Kluwer Academic Publishers, Dordrecht, pp. 163–177.
(1976) '*Terra Firma* and Infirma Species: From Medical Philosophical Anthropology to Philosophy of Medicine', *Journal of Medicine and Philosophy* 1(2), 104–135.
(1977) 'Medicine's Influence on Ethics: Reflections on the Putative Moral Role of Medicine', in S. F. Spicker and H. T. Engelhardt, Jr. (eds.), *Philosophical Medical*

Ethics: Its Nature and Significance, Kluwer Academic Publishers, Dordrecht, pp. 143–151.

(1977) 'Humanistic Studies in Medicine – A Report', in *Human Values Teaching Programs for Health Professionals* – 1976, Report no. 7, 3rd edition, Society for Health and Human Values, Institute on Human Values in Medicine, Philadelphia, pp. 33–45.

(1978) 'The Psychiatrist as Philosopher: Gewidmet Erwin W. Straus (October 11, 1891–May 20, 1975)', in H. T. Engelhardt, Jr. and S. F. Spicker (eds.), *Mental Health: Philosophical Perspectives*, Kluwer Academic Publishers, Dordrecht, pp. 144–155.

(1978) 'Gerontogentic Mentation: Memory, Dementia and Medicine in the Penultimate Years', in S. F. Spicker, Kathleen M. Woodward and David D. Van Tassel (eds.), *Aging and the Elderly: Humanistic Perspectives in Gerontology*, Humanities Press, Atlantic Highlands, New Jersey, pp. 153–180.

(1978) 'The Execution of Euthanasi: The Right of the Dying to a Re-Formed Health Care Context', in S. F. Spicker (ed.), *Organism, Medicine and Metaphysics: Essays in Honor of Hans Jonas on His 75th Birthday*, Kluwer Academic Publishers, Dordrecht, pp. 578–581.

(1978) 'The Role of Humanities in Geriatric Education', *The Gerontologist* 18(6), 578–581.

(1979) 'Perspectives on Death and Dying', *The Connecticut Scholar* 2, 43–63.

(1979) 'Clinical Judgment', in H. T. Engelhardt, Jr., S. F. Spicker, and B. Towers (eds.), *Clinical Judgment: A Critical Appraisal*, Kluwer Academic Publishers, Dordrecht, pp. 229–237.

(1979) 'Do We Have a Right to a Natural Death? Ethical, Legal and Moral Consideration', in S. F. Spicker (ed.), *Connecticut Medicine* 43 (October – Supplementary Issue), pp. 42–45.

(1981) 'The Bearing of Prognosis on the Ethics of Medicine: Congenital Anomalies, the Social Context and the Law', with J. R. Raye (co-author), in S. F. Spicker, J. M. Healey, and H. T. Engelhardt, Jr. (eds.), *The Law-Medicine Relations: A Philosophical Exploration*, Kluwer Academic Publishers, Dordrecht, pp. 189–216.

(1982) 'The Life-World and the Patient's Expectations of New Knowledge', in W. Bondeson, H. T. Engelhardt, Jr., S. F. Spicker, and J. White (eds.), *New Knowledge in the Biomedical Sciences: Some Moral Implications of Its Acquisition, Possession and Use*, Kluwer Academic Publishers, Dordrecht, pp. 205–215.

(1984) 'Chronicity: Some Definitional Issues', *Newsletter: Committee on Philosophy and Medicine, American Philosophical Association* (13), 2–3.

(1984) 'Person Ascriptions, Profound Disabilities, and Our Self-Imposed Duties', in L. Kopelman and J. C. Moskop (eds.), *Ethics and Mental Retardation*, Kluwer Academic Publishers, Dordrecht, pp. 25–26.

(1984) 'Comments on "On the Circle of Health"', in *Health, Disease, and Causal Explanations in Medicine*, Kluwer Academic Publishers, Dordrecht, pp. 25–26.

(1984) 'Philosophical Reflections on the "Biology of Aging"', in S. F. Spicker and S. Ingman (eds.), *Vitalizing Long-Term Care*, Springer Publishing Company, New York, pp. 29–45.

(1984) 'Philosophical Aspects of Brain Death', *Journal of Medicine and Philosophy* 9(4), 373–375.

(1984) 'Causes, Effects and Side Effects', with H. T. Engelhardt, Jr. (co-author), in L.

Nordenfelt and B. I. B. Lindahl (eds.), *Health, Disease, and Causal Explanation in Medicine*, Kluwer Academic Pubhlishers, Dordrecht, pp. 224–233.
(1984) 'The Propaganda of Science', *The Texas Humanist* 6(5), 16–18.
(1985) 'Coercion, Conversation, and the Casuist: A Reply to Jay Katz', in J. C. Moskop and L. Kopelman (eds.), *Ethics and Critical Care Medicine*, Kluwer Academic Publishers, Dordrecht, pp. 69–77.
(1986) 'Legal Aspects of Surrogate Motherhood are Evolving', *Norwich Bulletin*, B1–B2.
(1986) 'Ethical Reflections in Genetic Screening: A Reply to Swint and Kaback', in G. J. Agich and C. E. Begley (eds.), *The Price of Health*, Kluwer Academic Publishers, Dordrecht, pp. 157–164.
(1986) 'Aging: Two Features', *Human Values and Aging Newsletter* (Jan/Feb), 3–4.
(1986) 'Hospital Ethics Committees: Expanding Across the U.S.A.', in *Nikkei Medical*, McGraw Hill Publishing Co., p. 111.
(1986) 'Cognitive and Conative Issues in Contemporary Philosophy of Sciences', *Journal of Medicine and Philosophy* 11(1), 107–117. Translated into Chinese and reprinted in *Zhexue Yicong (Philosophical Translations)* 3 (1987), 73–78.
(1987) 'Selfish Safety or Redistributed Risks? Trade-Offs Among Automobile Occupants' Accident Fatality Risks', with H. C. Joksch and D. F. Mela (co-authors), in L. B. Lave, *Risk Assessment and Management*, Plenum Press, New York, pp. 261–277.
(1987) 'Experimentalism or Empiricism?', *Journal of Mental Imagery* 11(3), 1–2.
(1987) 'Contraception and Changing Images of Sexuality: Comments on Meredith Michaels' and Robert Solomon's Reflections', in S. F. Spicker, W. B. Bondeson and H. T. Engelhardt, Jr. (eds), *The Contraceptive Ethos: Redproductive Rights and Responsibilities*, Kluwer Academic Publishers, Dordrecht, pp. 241–248.
(1987) 'Max Delbruck's Mind from Matter? An Essay on Evolutionary Epistemology', *Journal of Biology and Philosophy* 2(4), 509–515.
(1987) 'In Defense of IEC', *Hastings Center Report* 17(1), 38–39.
(1987) 'An Introduction to the Medical Epistemology of Georges Canguilhem: Moving Beyond Michel Foucault', *Journal of Medicine and Philosophy* 12(4), 397–414.
(1987) 'Unfollowable Law: Comments on Nancy King's and Margery W. Shaw's Reflections', in R. C. McMillan, H. T. Engelhardt, Jr. and S. F. Spciker (eds.), *Euthanasia and the Newborn: Conflicts Regarding Saving Lives*, Kluwer Academic Publishers, Dordrecht, pp. 139–150.
(1987) 'L'epistemologie medicale francaise vue des U.S.A. (L'un et le Multiple)', *Prospective et Sante* ("Le Normal et le Pathologique") 40, (hiver), 53–59.
(1987) 'Prospective Payment: DRGs and Ethics', *Journal of Medicine and Philosphy* 12(2), 101–106.
(1987) 'Max Delbruck's Mind from Matter? An Essay on Evolutionary Epistemology', *Journal of Medicine and Philosophy* 12(3), 293–295.
(1988) 'Bochum Working Paper on the Practice of Medical Ethics', *Bochumer Arbeitsbogen Aur Medizinethischen Praxis*, Zentrum fur Medizinische Ethik Bochum, Medizinethische Materialien, Heft 2a, Bochum F.R.G., pp. 1–4.
(1988) 'Person Perception and the Death of the Person: A New Role for Health Professionals in Cases of Brain Death', in R. Zaner (ed.), *Death: Beyond Whole-Brain Criteria*, Kluwer Academic Publishers, Dordrecht, pp. 253–265.

(1988). 'Research Risks, Randomization, and Risks to Research: Reflections on the Produential Use of "Pilot" Trials', in S. F. Spicker, I. Alon, A de Vries and H. T. Engelhardt, Jr. (eds.), *The Use of Human Beings in Research: With Special Refenence to Clinical Trains*, Kluwer Academic Publishers, Dordrecht, pp. 1143–1160.

(1988) 'Rights, Reasonable, Exceptations, and Rationing: A Commentary on the Essays of Ruth Mattheis and Baruch Brody', in H.-M. Sass and R. Massey (eds.), *Health Care Systems: Moral Conflicts in European and American Public Policy*, Kluwer Academic Publishers, Dordrecht, pp. 237–253.

(1988) 'Marketing Health Care: Ethical Challenge to Physicians', in N. King, L. Churchill and A. Cross (eds.), *The Physician as Captain of the Ship: A Critical Reappraisal*, Kluwer Academic Publishers, Dordrecht, pp. 161–178.

(1988) 'Recht, Ansprache und Rationalisierungen im Gesundheitswesen', in H.-M. Sass (ed.), *Etik und Offentilishes Gesundheitswesen: Ordnungsethische und ordnungpolitische Einflussfaktoren im offeentlichen Gesundheitswesen*, Berlin, Heidelberg, Springer-Verlag, New York, pp. 65–78.

(1989) 'Comments on Barbara Starfield's "Child Health and Public Policy"', in L. M. Kopelman and J. C. Moskop (eds.), *Children and Health Care: Moral and Social Issues*, Kluwer Academic Publisher, Dordrecht, pp. 303–304.

(1989) 'Comments on John Ladd's "The Good Doctor and Medical Care of Chidlren"', in L. M. Kopelman and J. C. Moskop (eds.), *Children and Health Care: Moral and Social Issues*, Kluwer Academic Publisher, Dordrecht, pp. 23–26.

(1989) 'Access to Basic Health Care: A Moral Right or a Self-Imposed Duty?', *Quality Matters: A CHA Quality of Care Newsletter*, 4.

(1989) 'Ethical Considerations of the New Reproductive Technolgoies', in D. C. Byk (ed.), *Procreation artificelle ou en sont l'ethique et le droit? – contribution multidsiciplinaire et internationale*, Medicine at Societe, Editions Alexandre Lacassagne, Lyon, pp. 383–386.

(1989) 'Asclepias Shiden ni okeru Apollo no Yado America no Igaku-Rinri Kyoiku: sono Genjo to Kadai', A. Sakai (trans.), *Annals of the Japanese Association for Philosophical and Ethical Research in medicine*, pp. 113–122.

(1990) 'Ars Medicina at Conditio Humana: Edmund D. Pellegrino, M.D. on his 70th Birthday', with R. M. Ratzan (co-author), *Journal of Medicine and Philosophy* 15(3), 327–341.

(1990) 'Invulnerability and Medicine's "Promise" of Immortality: Changing Images of the Human Body During the Growth of Medical Knowledge', in H. ten have, G. Kimsma and S. F. Spicker (eds.), *The Growth of Medical Knowledge*, Kluwer Academic Publishers, Dordrecht, pp. 163–175.

(1990) 'Hospital Ethics Committees in the U.S.A.', *Nikkei Medical*, McGraw Hill Publishing Company 19(1), 40–43.

(1990) 'Medical Ethical Questionnaire in Diagnosis', Medizinethisch Meterialien, *Zentrum fur Medicinische Ethik*, Bochum, Heft 562, F.R.G., Mai, pp. 1–17.

(1990) 'Tasha no Kyoryoku niyoru Seishoku-Keikaku; Ho-teki, Tinri-tkei Kanten kara no Kosatsu', A. Sakai (trans.), *Annals of Japanese Association for Philosophical and Ethical Research in Medicine* 8, 90–101.

(1991) 'Entaku Togi', Y. Ito (trans.), in Atarashii Iryokan o Motomete, R. Ishiwata and A. Sakai (eds.), *Jiku Shuppan*,Tokyo, 171–173.

(1991) 'The Search for Bioethical Criteria to Select Renal Transplant Recipients', *Dialgoue: Canadian Philosophical Review* 30(3), 425–434.

(1991) 'Ethics in Diagnosis: Bodily Integrity, Trust-Telling, and the Good Physician', in J. L. Peset and D. Gracia (eds.), *The Ethics of Diagnosis*, Kluwer Academic Publishers, Dordrecht, pp. 107–122.

(1992) 'Chronicity: Some Definitional Issues', expanded version, *Ijigaku Knkyu (Journal of Interdisciplinary Humanistic Medical Research)*, The Society for Interdisciplinary Humanistic Medical Research (trans.), Iwate Medical University, Morioka, Japan.

(1992) 'Western Images of Mankind and Medicine', in K. Bai (ed.), *Towrad a New Replenishment of Medical Education and Hospital Service*, Kitasato University Press, Tokyo, Japan, pp. 252–278.

(1992) 'Marketing Health Care: Ethical Challenge to Physicians', in K. Bai (ed.), *Toward a New Replenishment of Medical Education and Hospital Service*, Kitasato University Press, Tokyo, Japan, pp. 336–345.

(1992) 'Healthcare Ethics Committees: Speical Problems with Consent', *Proceedings of a Conference on Ethical Decision-Making in Healthcare: Ethics Committees and Conflict Resolution Techniques*, Public Responsibility in Medicine and Research (PRIM&R), Boston, pp. 131–141.

(1992) 'Hodokoshi o Yamerukoto: Iryo Fukushi Kokka ni tsuite no Kasatsu to Rinriteki Kanten kara no Hihan', A. Sakai (trans.), *Journal of Health Care, Medicine and Community* 2, 15–26.

(1993) 'Intuition and the Process of Medical Diagnosis: The Quest for Explicity Knowledge in the Technological Era', in C. Delkeskamp-Hayes and M. A. Cutter (eds.), Science, *Technology, and the Art of Medicine: European-American Dialogues*, Kluwer Academic Publishers, Dordrecht, pp. 199–210.

(1993) 'Going off the Dole: A Prudential and Ethical Critique of the "Healthfare" State', *Health Care Analysis* 1(1), pp. 33–38. Reprinted in D. Seedhouse (ed.), *Reforming Health Care; The Philosophy and Practice of International Health Reform*, John Wiley and Sons, Ltd., 1995, pp. 131–141.

(1993) 'Op eigen benen leren staan: Een perspectief', J. Welie (trans.), *Tijdschrift voor Geneeskunde en Ethiek*.

(1994) 'Aging and Dependency in a Prosthetic Age: Humanity in the Penultimate State of Embodiment', *The Southwest Journal on Aging* 9(2), 41–49.

(1994) 'ICU Triage: The Ethics of Scarcity, the Ideal of Impartiality and the Inadvertent Endorsement of Evil', in R. Lee and D. Morgan (eds.), *Death Rites: Law and Ethics at the End of Life*, Routledge, Ltd., London/New York, pp. 91–101.

(1994) 'A Proposal to Establish the Office of Healthcare Education in Ethics and Law (HEEAL) in U.D. Hospitals', C. Zuckerman, *HEC Forum* 6(3), 126–138.

(1995) 'The Language of Aging Continues to Bewitch Us', *The Southwest Journal on Aging* 10.

(1996) 'Government and Bureaucratic Bioethics: Addressing Moral Issues in the Service of Ideology', *Journal of Medicine and Philosophy* 21(2), 113–119.

NOTES ON CONTRIBUTORS

George J. Agich, Ph.D., Professor, Department of Bioethics, The Cleveland Clinic Foundation, Cleveland, OH, U.S.A.

Thomas J. Bole, Ph.D., Ph.D., Holy Archangels Monastery, Kendalia, Texas, U.S.A.

Mark J. Cherry, Ph.D., Assistant Professor, Department of Philosophy, Saint Edward's University, Austin, Texas, U.S.A.

Bernard M. Dickens, Ph.D., LL.D., Professor of Law, Faculty of Law, Faculty of Medicine and Joint Centre for Bioethics, University of Toronto, Ontario, Canada.

H. Tristram Engelhardt, Jr., Ph.D., M.D., Professor, Center for Medical Ethics and Health Policy, Baylor College of Medicine; and Department of Philosophy, Rice University, Houston, Texas, U.S.A.

K. W. M. Fulford, Ph.D., Professor of Philosophy and Mental Health, University of Warwick, and Honorary Consultant Psychiatrist, University of Oxford, Oxford, United Kingdom.

Thomas Halper, Ph.D., Professor, Chairman, Department of Political Science, Baruch College and the City University of New York Graduate School New York, New York, U.S.A.

Loretta M. Kopelmann, Ph.D., Professor, Chairman, Department of Medical Humanities, East Carolina School of Medicine, Greenville, North Carolina, U.S.A.

Akio Sakai, M.D., Professor, School of Medicine, Iwate Medical University, Morioka, Japan.

H. T. Engelhardt, Jr. (ed.), The Philosophy of Medicine, 241–242.

Henk A. M. J. ten Have, M.D., Ph.D., Professor, Faculty of Medical Sciences and Department of Ethics, Philosophy and History of Medicine, Katholieke Universiteit Nijmegen, Nijmegen, The Netherlands.

Marx W. Wartofsky, Ph.D., Formerly Professor, Department of Philosophy, Baruch College, City University of New York, New York, New York, U.S.A. Deceased 1997.

INDEX

Philosophy and Medicine

1. H. Tristram Engelhardt, Jr. and S.F. Spicker (eds.): *Evaluation and Explanation in the Biomedical Sciences.* 1975 ISBN 90-277-0553-4
2. S.F. Spicker and H. Tristram Engelhardt, Jr. (eds.): *Philosophical Dimensions of the Neuro-Medical Sciences.* 1976 ISBN 90-277-0672-7
3. S.F. Spicker and H. Tristram Engelhardt, Jr. (eds.): *Philosophical Medical Ethics.* Its Nature and Significance. 1977 ISBN 90-277-0772-3
4. H. Tristram Engelhardt, Jr. and S.F. Spicker (eds.): *Mental Health.* Philosophical Perspectives. 1978 ISBN 90-277-0828-2
5. B.A. Brody and H. Tristram Engelhardt, Jr. (eds.): *Mental Illness.* Law and Public Policy. 1980 ISBN 90-277-1057-0
6. H. Tristram Engelhardt, Jr., S.F. Spicker and B. Towers (eds.): *Clinical Judgment.* A Critical Appraisal. 1979 ISBN 90-277-0952-1
7. S.F. Spicker (ed.): *Organism, Medicine, and Metaphysics.* Essays in Honor of Hans Jonas on His 75th Birthday. 1978 ISBN 90-277-0823-1
8. E.E. Shelp (ed.): *Justice and Health Care.* 1981 ISBN 90-277-1207-7; Pb 90-277-1251-4
9. S.F. Spicker, J.M. Healey, Jr. and H. Tristram Engelhardt, Jr. (eds.): *The Law-Medicine Relation.* A Philosophical Exploration. 1981 ISBN 90-277-1217-4
10. W.B. Bondeson, H. Tristram Engelhardt, Jr., S.F. Spicker and J.M. White, Jr. (eds.): *New Knowledge in the Biomedical Sciences.* Some Moral Implications of Its Acquisition, Possession, and Use. 1982 ISBN 90-277-1319-7
11. E.E. Shelp (ed.): *Beneficence and Health Care.* 1982 ISBN 90-277-1377-4
12. G.J. Agich (ed.): *Responsibility in Health Care.* 1982 ISBN 90-277-1417-7
13. W.B. Bondeson, H. Tristram Engelhardt, Jr., S.F. Spicker and D.H. Winship: *Abortion and the Status of the Fetus.* 2nd printing, 1984 ISBN 90-277-1493-2
14. E.E. Shelp (ed.): *The Clinical Encounter.* The Moral Fabric of the Patient-Physician Relationship. 1983 ISBN 90-277-1593-9
15. L. Kopelman and J.C. Moskop (eds.): *Ethics and Mental Retardation.* 1984 ISBN 90-277-1630-7
16. L. Nordenfelt and B.I.B. Lindahl (eds.): *Health, Disease, and Causal Explanations in Medicine.* 1984 ISBN 90-277-1660-9
17. E.E. Shelp (ed.): *Virtue and Medicine.* Explorations in the Character of Medicine. 1985 ISBN 90-277-1808-3
18. P. Carrick: *Medical Ethics in Antiquity.* Philosophical Perspectives on Abortion and Euthanasia. 1985 ISBN 90-277-1825-3; Pb 90-277-1915-2
19. J.C. Moskop and L. Kopelman (eds.): *Ethics and Critical Care Medicine.* 1985 ISBN 90-277-1820-2
20. E.E. Shelp (ed.): *Theology and Bioethics.* Exploring the Foundations and Frontiers. 1985 ISBN 90-277-1857-1

Philosophy and Medicine

21. G.J. Agich and C.E. Begley (eds.): *The Price of Health.* 1986
ISBN 90-277-2285-4
22. E.E. Shelp (ed.): *Sexuality and Medicine.* Vol. I: Conceptual Roots. 1987
ISBN 90-277-2290-0; Pb 90-277-2386-9
23. E.E. Shelp (ed.): *Sexuality and Medicine.* Vol. II: Ethical Viewpoints in Transition. 1987 ISBN 1-55608-013-1; Pb 1-55608-016-6
24. R.C. McMillan, H. Tristram Engelhardt, Jr., and S.F. Spicker (eds.): *Euthanasia and the Newborn.* Conflicts Regarding Saving Lives. 1987
ISBN 90-277-2299-4; Pb 1-55608-039-5
25. S.F. Spicker, S.R. Ingman and I.R. Lawson (eds.): *Ethical Dimensions of Geriatric Care.* Value Conflicts for the 21th Century. 1987 ISBN 1-55608-027-1
26. L. Nordenfelt: *On the Nature of Health.* An Action-Theoretic Approach. 2nd, rev. ed. 1995 SBN 0-7923-3369-1; Pb 0-7923-3470-1
27. S.F. Spicker, W.B. Bondeson and H. Tristram Engelhardt, Jr. (eds.): *The Contraceptive Ethos.* Reproductive Rights and Responsibilities. 1987
ISBN 1-55608-035-2
28. S.F. Spicker, I. Alon, A. de Vries and H. Tristram Engelhardt, Jr. (eds.): *The Use of Human Beings in Research.* With Special Reference to Clinical Trials. 1988
ISBN 1-55608-043-3
29. N.M.P. King, L.R. Churchill and A.W. Cross (eds.): *The Physician as Captain of the Ship.* A Critical Reappraisal. 1988 ISBN 1-55608-044-1
30. H.-M. Sass and R.U. Massey (eds.): *Health Care Systems.* Moral Conflicts in European and American Public Policy. 1988 ISBN 1-55608-045-X
31. R.M. Zaner (ed.): *Death: Beyond Whole-Brain Criteria.* 1988
ISBN 1-55608-053-0
32. B.A. Brody (ed.): *Moral Theory and Moral Judgments in Medical Ethics.* 1988
ISBN 1-55608-060-3
33. L.M. Kopelman and J.C. Moskop (eds.): *Children and Health Care.* Moral and Social Issues. 1989 ISBN 1-55608-078-6
34. E.D. Pellegrino, J.P. Langan and J. Collins Harvey (eds.): *Catholic Perspectives on Medical Morals.* Foundational Issues. 1989 ISBN 1-55608-083-2
35. B.A. Brody (ed.): *Suicide and Euthanasia.* Historical and Contemporary Themes. 1989 ISBN 0-7923-0106-4
36. H.A.M.J. ten Have, G.K. Kimsma and S.F. Spicker (eds.): *The Growth of Medical Knowledge.* 1990 ISBN 0-7923-0736-4
37. I. Löwy (ed.): *The Polish School of Philosophy of Medicine.* From Tytus Chałubiński (1820–1889) to Ludwik Fleck (1896–1961). 1990
ISBN 0-7923-0958-8
38. T.J. Bole III and W.B. Bondeson: *Rights to Health Care.* 1991
ISBN 0-7923-1137-X

Philosophy and Medicine

39. M.A.G. Cutter and E.E. Shelp (eds.): *Competency.* A Study of Informal Competency Determinations in Primary Care. 1991 ISBN 0-7923-1304-6
40. J.L. Peset and D. Gracia (eds.): *The Ethics of Diagnosis.* 1992 ISBN 0-7923-1544-8
41. K.W. Wildes, S.J., F. Abel, S.J. and J.C. Harvey (eds.): *Birth, Suffering, and Death.* Catholic Perspectives at the Edges of Life. 1992 [CSiB-1] ISBN 0-7923-1547-2; Pb 0-7923-2545-1
42. S.K. Toombs: *The Meaning of Illness.* A Phenomenological Account of the Different Perspectives of Physician and Patient. 1992 ISBN 0-7923-1570-7; Pb 0-7923-2443-9
43. D. Leder (ed.): *The Body in Medical Thought and Practice.* 1992 ISBN 0-7923-1657-6
44. C. Delkeskamp-Hayes and M.A.G. Cutter (eds.): *Science, Technology, and the Art of Medicine.* European-American Dialogues. 1993 ISBN 0-7923-1869-2
45. R. Baker, D. Porter and R. Porter (eds.): *The Codification of Medical Morality.* Historical and Philosophical Studies of the Formalization of Western Medical Morality in the 18th and 19th Centuries, Volume One: Medical Ethics and Etiquette in the 18th Century. 1993 ISBN 0-7923-1921-4
46. K. Bayertz (ed.): *The Concept of Moral Consensus.* The Case of Technological Interventions in Human Reproduction. 1994 ISBN 0-7923-2615-6
47. L. Nordenfelt (ed.): *Concepts and Measurement of Quality of Life in Health Care.* 1994 [ESiP-1] ISBN 0-7923-2824-8
48. R. Baker and M.A. Strosberg (eds.) with the assistance of J. Bynum: *Legislating Medical Ethics.* A Study of the New York State Do-Not-Resuscitate Law. 1995 ISBN 0-7923-2995-3
49. R. Baker (ed.): *The Codification of Medical Morality.* Historical and Philosophical Studies of the Formalization of Western Morality in the 18th and 19th Centuries, Volume Two: Anglo-American Medical Ethics and Medical Jurisprudence in the 19th Century. 1995 ISBN 0-7923-3528-7; Pb 0-7923-3529-5
50. R.A. Carson and C.R. Burns (eds.): *Philosophy of Medicine and Bioethics.* A Twenty-Year Retrospective and Critical Appraisal. 1997 ISBN 0-7923-3545-7
51. K.W. Wildes, S.J. (ed.): *Critical Choices and Critical Care.* Catholic Perspectives on Allocating Resources in Intensive Care Medicine. 1995 [CSiB-2] ISBN 0-7923-3382-9
52. K. Bayertz (ed.): *Sanctity of Life and Human Dignity.* 1996 ISBN 0-7923-3739-5
53. Kevin Wm. Wildes, S.J. (ed.): *Infertility: A Crossroad of Faith, Medicine, and Technology.* 1996 ISBN 0-7923-4061-2
54. Kazumasa Hoshino (ed.): *Japanese and Western Bioethics.* Studies in Moral Diversity. 1996 ISBN 0-7923-4112-0

Philosophy and Medicine

55. E. Agius and S. Busuttil (eds.): *Germ-Line Intervention and our Responsibilities to Future Generations.* 1998 ISBN 0-7923-4828-1
56. L.B. McCullough: *John Gregory and the Invention of Professional Medical Ethics and the Professional Medical Ethics and the Profession of Medicine.* 1998 ISBN 0-7923-4917-2
57. L.B. McCullough: *John Gregory's Writing on Medical Ethics and Philosophy of Medicine.* 1998 [CiME-1] ISBN 0-7923-5000-6
58. H.A.M.J. ten Have and H.-M. Sass (eds.): *Consensus Formation in Healthcare Ethics.* 1998 [ESiP-2] ISBN 0-7923-4944-X
59. H.A.M.J. ten Have and J.V.M. Welie (eds.): *Ownership of the Human Body.* Philosophical Considerations on the Use of the Human Body and its Parts in Healthcare. 1998 [ESiP-3] ISBN 0-7923-5150-9
60. M.J. Cherry (ed.): *Persons and Their Bodies.* Rights, Responsibilities, Relationships. 1999 ISBN 0-7923-5701-9
61. R. Fan (ed.): *Confucian Bioethics.* 1999 [APSiB-1] ISBN 0-7923-5853-8
62. L.M. Kopelman (ed.): *Building Bioethics.* Conversations with Clouser and Friends on Medical Ethics. 1999 ISBN 0-7923-5853-8
63. W.E. Stempsey: *Disease and Diagnosis.* 2000 PB ISBN 0-7923-6322-1
64. H.T. Engelhardt (ed.): *The Philosophy of Medicine.* Framing the Field. 2000 ISBN 0-7923-6223-3
65. S. Wear, J.J. Bono, G. Logue and A. McEvoy (eds.): *Ethical Issues in Health Care on the Frontiers of the Twenty-First Century.* 2000 ISBN 0-7923-6277-2

KLUWER ACADEMIC PUBLISHERS – DORDRECHT / BOSTON / LONDON

Lightning Source UK Ltd.
Milton Keynes UK
UKOW030753271211

184386UK00001B/4/P

9 789048 154197